Nurturing Inquiry

Nurturing Inquiry

Real Science for the Elementary Classroom

CHARLES R. PEARCE

FOREWORD BY WENDY SAUL

HEINEMANN
Portsmouth, NH

Heinemann
A division of Reed Elsevier Inc.
361 Hanover Street
Portsmouth, NH 03801-3912
http://www.heinemann.com

Offices and agents throughout the world

The author and publisher thank those who generously gave permission to reprint borrowed material.
Figures 4-1 and 4-3 reprinted by permission of Charles Pearce. In *Science Workshop: A Whole Language Approach*, edited by Wendy Saul, et al. (Heinemann, A division of Reed Elsevier Inc., Portsmouth, NH, 1993).

Library of Congress Cataloging-in-Publication Data
Pearce, Charles R.
Nurturing inquiry : real science for the elementary classroom / Charles R. Pearce : foreword by Wendy Saul.
p. cm.
Includes bibliographical references and index.
ISBN 0-325-00135-9
1. Science—Study and teaching (Elementary) 2. Science—Methodology. I. Title.
LB1584.P34 1999
372.3'5'044—dc21 98-52475
CIP

Editor: Hilary Breed Van Dusen
Production: Melissa L. Inglis
Cover design: Darci Mehall/Aureo Design
Manufacturing: Louise Richardson

Printed in the United States of America on acid-free paper
03 02 01 00 99 RRD 1 2 3 4 5

Contents

Foreword

The first time I visited Charles Pearce's classroom, I sat in the back of the room, struggling to take notes. I faced two basic problems. First, how could I capture in words the mystery and excitement and pleasure of learning the children obviously felt? The other problem? I wanted to simply put away my scholarly tools, slip into my fifth-grade self, and join the class as a student. I felt like I was watching a maestro conduct a symphony, and I wanted to take out my oboe and play.

Nurturing Inquiry is an important book for educators who, like Mr. Pearce, view teaching as creative, intellectually engaging, and exceedingly interesting work. In rich, concrete detail, the author describes both the activity and the decision making that goes on in his classroom. In doing so he provides us with a volume that not only gives us the how-tos of getting inquiry established, but also models the reflection, choices, and values that shape the practice of a highly successful teacher. *Nurturing Inquiry* is a book about the everyday process of teaching as well as a guide to helping others think about their own challenges and opportunities.

There has been much talk in recent years about the need to recognize teacher wisdom and excellent practice. Excellent practitioners are often asked to participate in reform efforts or curriculum-evaluation groups, and most do so willingly. And then when the final document is released these teachers look at the collected responses to a problem—the new curriculum, the report on reform—and wonder what happened to their ideas. Somehow, in the process of being digested and turned into the language of goals and objectives, we lose the picture of what actually happens in a successful classroom: How does the day begin? How does the year evolve? How does the thoughtful practitioner evaluate him or herself?

This book, by contrast, offers teachers an individualized voice. Instead of sanitized, teacher-proof directives, Charles Pearce shows us in his distinctive, well-defined way that teachers are capable of intelligence and good judgment. Teaching is still too often seen as a career for those who can't do anything else. The existence of *Nurturing Inquiry* serves not only to defy this stereotype, but also to say to teachers everywhere, "Look what is possible. See yourselves as part of a community of educators who are smart and capable. Join in by viewing your own practice as critically important and original."

What characterizes the work of teachers like Charles Pearce, teachers in this community of highly able practitioners? First and foremost, there is a belief that the job of the educator is to help children recognize, explore, and celebrate logical, rigorous thought and elegant reasoning. And to promote this kind of thinking, this inquiry, the teacher's most powerful tool is a vibrant trust in children—that they are smart, that they will "get it," that there is pleasure to be had in "figuring things out." Unlike some of us who can barely keep from stepping in to show young learners the way, Charlie lets us see, from his perspective, the power and pleasure of standing back.

But teaching, as it unfolds in this book, is not simply a matter of watching. Pearce shows us how to scaffold experiences for young people, provide them with playful but meaningful challenges, and create opportunities for them to worry ideas into place. In this book we also get some sense of this teacher's genuine interest in the observations and explanations his students offer. Pearce unravels the formulaic conversation on the teaching of thinking and replaces it with authentic methods and attitudes, the habits of mind employed by practicing scientists.

Typically, scientific thinking is tied to something called "the scientific method." Part of what I love about this book is that Pearce really understands scientific method, not just as a series of steps beginning with a hypothesis and ending with a conclusion, but rather as a glorious, rocky trail, strewn with diversions that may, in fact, result in more interesting questions and conclusions than the mapped-out path. Sometimes students need a guide to encourage them down this path, to help them look around. At other times the guide serves as a distraction and the "good teacher" simply helps children read nature's signs. Together they discover what paths seem to lead somewhere and which result in dead ends. Together they cope with impediments. Together they backtrack. How far back should a student go? The answers aren't always clear. They are not even always interesting, but when a person is really trying to figure something out, that hardly seems to matter. This is the scientific method, which authentic scientists practice. This is the method that Pearce teaches his students. And these are the methods he shares with us, his readers.

Nurturing Inquiry helps us tour Mr. Pearce's classroom from our reading chairs. The content of the book is rich and varied and the explanations of process give any practitioner a sense of "what's next." But I hope that in absorbing this volume, readers learn as much about the tone of the activity that takes place as they do about the content. This is a book that documents the intellectual and attitudinal growth of children. It is also a book about the growth of a teacher, a teacher who continues to learn and explore with the same inquisitiveness he seeks to instill in children. For me it is both accurate and inspiring.

But I am also aware that the inquisitiveness and humility that drive Pearce's teaching and writing mean that his practice will change. I have no doubt that if we were to visit this classroom three or four years from now, we would see merely vestiges of the original ideas and activities addressed here. Their evolution would have been in direct response to the interests and experiences of the children in his class. His strategies and methods would also have been affected by the material and physical resources with which he is surrounded—a year of warm weather, a gift of petri dishes, the assignment of a student teacher with a love for art. His classroom may have also changed as a result of the response of other practitioners. For Charles Pearce knows, as any good teacher knows, that one of the best ways to improve upon your practice is to listen to what others have done with it. Read *Nurturing Inquiry* not only for the score, but for the nuance, the energy, the underlying sense of composition that guides the conductor. And then think of your own class. What possibilities await those who teach!

WENDY SAUL

Acknowledgments

Writing *Nurturing Inquiry* has been a process of discovery and documentation. Like science, writing begins with questions: What can I share? How can I capture in words the essence of inquiry? Throughout my work of implementing and documenting inquiry science in my classroom, many people have offered encouragement, ideas, and friendship.

Nurturing Inquiry began with my participation in the Elementary Science Integration Project at the University of Maryland, Baltimore County (UMBC). There, Dr. Wendy Saul shared her inspired—and inspiring—belief that teachers can forge real links between science, reading, and writing when they engage in authentic activities. Other colleagues at ESIP helped to make the first Kids' Inquiry Conferences a reality. As ESIP program director, Barbara Bourne brought a gift for being practical and creative at the same time. Her ideas about logistics made it possible for KIC to be replicated in many places by many teachers.

Once inspired, however, a teacher needs freedom to try new ideas. In the schools of Carroll County, Maryland, I found principals and supervisors who supported my efforts. Bonnie Ferrier, my first principal, encouraged risk taking and enthusiastically believed in inquiry science. In Larry Tyree I found a supervisor who recognized that objectives can be met in different ways. Bob Mitchell, principal of Manchester Elementary, continues to help make the Kids' Inquiry Conference possible for my students and others, supporting students and teachers as they strive to grow.

My own children, Will, Emily, and Sarah, let me watch them be scientists—spinning grapes on a plate, driving cars down ramps, wondering where candy canes disappear to when they are eaten. Watching them prepared me to give the same freedom and encouragement to my student scientists at school. I learned a lot from the children I have taught, and I thank them for all the moments of discovery.

To Hilary Breed Van Dusen, my editor at Heinemann, I owe my thanks for her patience and expertise. We may have established a new record for e-mail correspondence.

Finally, I wish to thank my wife, Karen, from whom I have learned much. Her belief in me and in this project sustained me through the long hours of dueling with a sometimes reluctant computer, searching through accumulated papers for just the right example, and trying to put the wonder of teaching science here on these pages. For her encouragement and assistance, which has made *Nurturing Inquiry* a reality, I will forever be grateful.

Introduction

I have always wondered why kids like science. What is it about science that kids come to school wanting to do? There are the obvious reasons for their interest: a diversion in the school day, experiments to be performed. But what about the science that kids do at home? What drives them there?

In their play, children encounter science all of the time. Bouncing a ball, flying a kite, observing an insect—these are activities that are scientific in nature. But where is the line drawn between play and work, at home or at school? What motivates children (and ourselves) to do either? Recently I had a conversation with a gentleman who was lamenting the elimination of cash awards for elementary science fair winners. "Where is the motivation?" he asked. "Why will the students want to do science without a prize?" He found it hard to believe that children don't do science for financial gain.

Kids at the beach are not investigating sand and waves to earn a trophy from the lifeguard. Children rolling grapes on a paper plate or looking under rocks in a stream don't care about adults in suits making judgments on their investigations. "Rewards" invented by adults to separate winners from losers do not create scientists.

Children are authentically motivated to do science for one basic reason: to find out! Kids are intrigued by the unknown. The world around them is a mystery to be unraveled and solved. Place any child in any place and he or she will begin to explore. Ideas will develop, questions will flow, discoveries will be made. Although it may drive parents and teachers crazy at times, it is not natural for kids to simply sit passively and do nothing. Every one of our students is driven to actively find things out, one way or another.

It is this innate drive that has made our species so successful. And it is this trait in our students that can be utilized to help every child succeed in school. The children whose stories are told in this book won no prizes. They were not motivated by ribbons or certificates. They simply were given the time, materials, and opportunity to find things out in a structured environment.

The national science standards identify several different sets of standards for science education. First among the content standards is "science as inquiry." Inquiry is an amazing equalizer in the classroom. Children at all levels are able to shine as they assume the autonomy (and responsibility) of inquiry. Frequently the same kids who struggle to succeed in school show the others with whom they are working what they have discovered and learned. Every child is gifted in some area. Our responsibility is to provide the opportunities that will enhance student success, self-esteem, and a desire to communicate. An inquiry approach to science instruction can accomplish this goal.

It was through inquiry that each of us began learning as toddlers. It is through inquiry that we as adults continue to learn. This book is about providing those kinds of opportunities for our children in their classrooms.

SECTION 1

Lighting the Fire: Getting Inquiry Under Way

One

Inquiry: The Next Frontier

Every child is a scientist. Children think in ways that scientists think, say things that scientists say, and do things that scientists do. What pure science it is when a child touches and feels, tastes and senses, examines and manipulates. Children are driven to fully experience all they can in their surroundings. As a result, the students with whom we teach possess rich backgrounds of experience and vast databases of information. Visits to the playpen, sandbox, basement, corner lot, backyard, or playground are visits to laboratories in which astounding discoveries are made every day.

Our species is inherently scientific. For nearly two million years we have been observing, classifying, measuring, collecting data, predicting, and communicating. Our survival has depended not so much on size or strength or speed but on our ability to do science; to know as much as we can about our environment and then use that information to adapt and create and prosper.

What an awesome task it is, then, to be called upon to teach science. What can teachers in our schools give experienced scientists who have spent their lifetimes acquiring knowledge through their own investigations and discoveries? How can we nurture their innate curiosity without inhibiting its growth? What experiences can we provide that might enhance the development of the scientist within each of the children we teach?

From infancy on, children learn on their own; they learn by doing. They explore, touch, take apart, listen, examine, infer, relate, connect, extrapolate, create, destroy, and conclude until stopped by fatigue or an exasperated parent who has finally lost patience. Without being taught the steps, children use the processes of science. The best most parents can do is simply keep their children from hurting themselves as the investigations become ever more ambitious. We don't force kids to learn. The act of play is in itself an intense scientific study, unassigned and internally motivated. Science may well be the only content area for which all children come to school prepared. Yet, it seems that the traditional style of teaching in our schools differs greatly from the child's mode of learning elsewhere.

Not so many years ago, the widely accepted method of science instruction was by following a textbook with end-of-chapter questions. This approach was orderly, well-controlled, and easy to assess (although assessment was of dubious value). Science, on the other hand, has never been orderly or well-controlled or easy to assess. Moreover, away from school, kids learn in ways on their own that are anything *but* orderly or controlled. Textbook instruction fell short in several ways; students learned facts without process, and were not permitted to use their own styles of learning. Science instruction had to include more.

Teacher demonstrations in science class have always been fun and exciting. Rather than reading about scientific experiments in a textbook, students witness the teacher using science equipment to prove a concept. When utilized in addition to a text, teacher demonstrations bring science alive in the classroom. For the teacher, demonstrations are orderly, well-controlled, and independent. For the students, however, it can be frustrating to watch the teacher have all the fun. Teacher demonstrations may have their place in the classroom, but science instruction must include more.

Hands-on science moved the science materials from the teacher's table to the students' desks. At last, with the hands-on philosophy spreading among educators, children have been able to touch, manipulate, and see close up many of the amazing things previously reserved for the teacher herself. With carefully planned directions and specially selected materials, hands-on science offered a cookbook approach. The teacher has been provided with step-by-step procedures to lead the students through an array of activities. Though not as orderly as textbooks or teacher demonstrations, hands-on practice is well-controlled and relatively easy to assess with clearly defined right and wrong answers and results. (Hands-on programs contain specific outcomes and results that the children are expected to successfully replicate.) A hands-on approach toward science instruction would seem to be the ideal method of teaching science. When compared with reading textbooks or watching teacher demonstrations, hands-on science seems to be an improvement that has become the preferred means of science instruction in many classrooms. But should science instruction include more?

A hands-on approach in the classroom is still vastly different from the way children learn on their own. In the sandbox, a child does not follow directions or use only those materials specified for a particular activity. There are no predetermined results that must be replicated to be correct. The child in the sandbox is not restricted by someone else's agenda. When sifting is completed, the sifter is not put away, never to be seen again. Upon examination, the traditional hands-on method of learning science is anything *but* the way kids do science on their own.

Perhaps every classroom has a place for up-to-date science textbooks when used as reference tools, and a place for teacher demonstrations when materials are in short supply or unsafe for children to handle. And of course, each classroom has a place for hands-on science to help children properly use materials and acquire prior knowledge. But for of a class of experienced child-scientists, is this enough to fully develop that inner drive to investigate and discover? What, then, is the something more that might be provided?

As children, each of us learned on our own through the process of inquiry. We asked questions and had our own problems to solve. We developed experiments and investigations to answer those questions. We gathered data, observed results, and drew conclusions. It could even be said that we were using the scientific method without knowing the terms for the steps. Bringing this inquiry approach to science is the next step in science instruction.

Inquiry science in the classroom helps teachers to meet the students where they are when they come to class. Inquiry provides authenticity and autonomy. It capitalizes on textbooks, teacher demonstrations, and hands-on activities. Furthermore, this approach affords the students opportunities to do what kids do best: investigate, explore, and discover, using their own questions, curiosities, and interests. An inquiry approach is not orderly, well-controlled, or easy to assess, but neither is science itself. The value of inquiry science is that it permits children to continue to develop the science skills that began in them at birth.

As I first began thinking about providing experiences in my class that paralleled the way kids learn naturally through inquiry, I was concerned with a number of troubling questions. How could I provide an authentic environment of inquiry in a class of thirty students? Where would the materials come from? How could they be managed? What about accountability? How could I be assured that precious school time would be used wisely? But with these questions came some intriguing possibilities. Could this inquiry approach enhance other areas of the curriculum, especially language arts and math? Might this approach help students to become self-motivated at school as they had always been in their out-of-school investigations? Could an inquiry approach to science create a community of scientists in our classrooms?

Implementing inquiry science in the classroom is not easy. The children are often going in divergent directions. Many are interested in questions and discoveries far more complex than any the science curriculum writers had envisioned. One group may be investigating bacteria growth while another designs a structure to withstand earthquakes. Others are writing articles or reading books or using the computer for research. Children are learning from one another, preparing presentations to deliver at scientific conferences and carefully documenting their own discoveries. For the teacher, providing background information, materials, and time for all this activity is both challenging and exhilarating. Inquiry science takes perseverance by both the teacher and students, but it is rewarding for everyone.

The inquiry model (Figure 1-1) conceptualizes some of the many components of an inquiry approach to science. The process begins with the students: their interests, curiosities, and experiences. From the students come questions and ideas for investigation. Then, as discoveries are made and documented, the student scientists report to one another through a variety of forms. The cycle of inquiry can spiral back toward the child herself to begin again, or it can lead to other pursuits, inspiring children to carry on what was begun or to make innovations in new areas. Woven

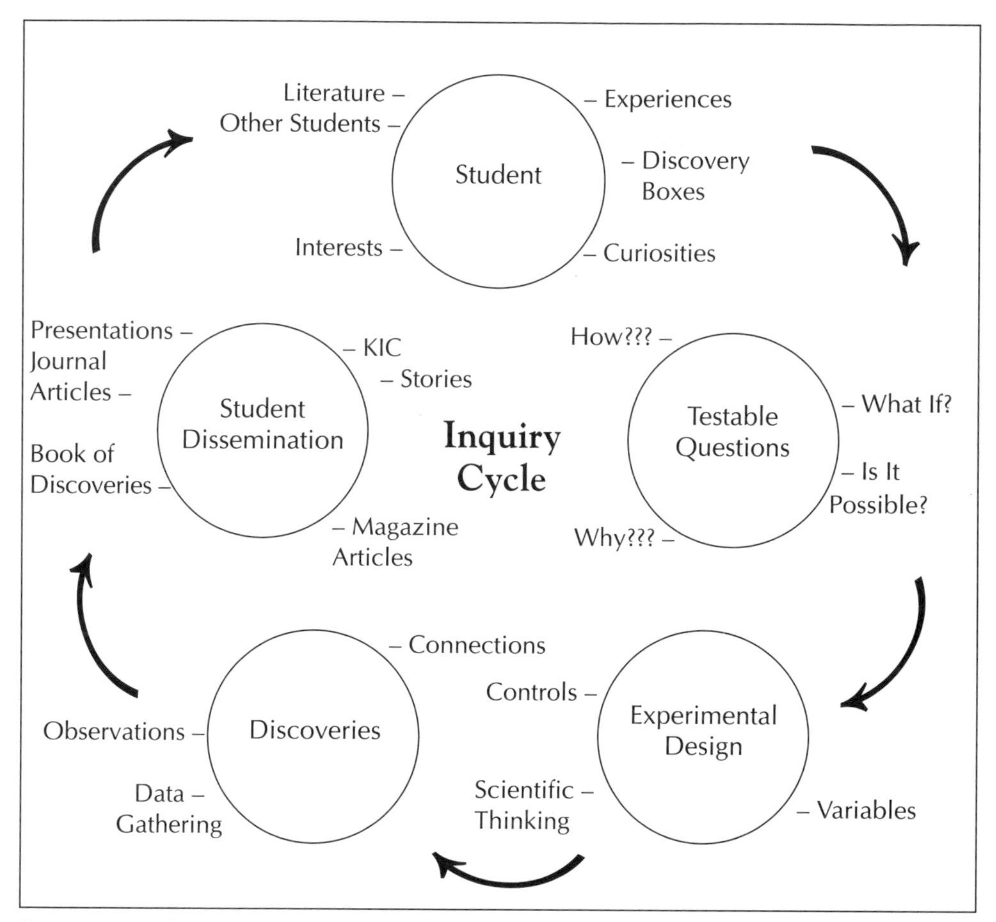

FIGURE 1-1 *The Inquiry Cycle*

throughout the cycle is authenticity. Not only does it show how *children* think about and do science, but it also shows what may occur in the overall scientific community.

Throughout this book we will explore how inquiry can be incorporated in the daily activities of any classroom. The components of the inquiry model fit together in a spiral, each building on the other as inquiry skills grow within each child. This model represents several years of evolution in a classroom, where it continues to develop. Its form can be adapted and changed to suit many needs.

There is no single correct method to incorporate inquiry into a classroom. Providing kids with the autonomy of inquiry can be accomplished in a variety of ways. Portrayed here is one model that has brought success with students of differing ability levels and interests.

Inviting inquiry science into the classroom does not mean chaos will follow. For inquiry to be manageable and beneficial, structure must be planned and implemented. Formal activities at the start of the year will give way to less controlled and restricted activities later. Trusting each student as a scientist also means expecting clearly defined outcomes. Students must understand that as scientists they are expected to question and probe; to design fair tests; to gather data and clearly record discoveries; to compare findings with one another as they search for patterns and connections; and to assess the validity of their own research and the research of others. The following chapters describe a philosophy and range of ideas that will help guide students to inquire from the first day of the school year and help sustain inquiry to the very last.

Inquiry in Action: Life on Mars?

On July 4, 1997 the Mars Pathfinder spacecraft touched down on the red planet. As it deployed and began sending back pictures and data, people around the world marveled at the achievement, and the scientists involved were elated. Questions about Mars were about to be answered and so many new questions were about to be asked.

One of the early discoveries from the data was that water had at one time been in abundant supply on the surface of Mars where the Pathfinder had landed. The evidence seemed clear that Mars had at one time been quite similar to Earth and it still shared many characteristics. If this were so, the scientists asked, could life have possibly existed on Mars eons ago? And if it had, what evidence of life might be present to prove its existence on Mars?

These kinds of questions reminded me of research presented five years earlier at a scientific conference. Two scientists had made an astounding hypothesis. Having made and studied erosion models, these scientists noticed a striking similarity between dry riverbeds in their models and photographs of the surface of Mars. In those photographs there appeared to be evidence of ancient rivers. These scientists had also been researching rocks. They knew that sedimentary rocks required water in order to form. They also knew that fossils, when found, are almost always located within the layers of sedimentary rocks.

The two scientists speculated that if we went to Mars and landed in a delta or dry riverbed, we would find sedimentary rocks. If there had ever been life on Mars, they told us, its evidence might be hidden within the layers of those rocks.

The scientists reporting were fifth-grade students at a science conference for children. At the time, based upon data from the Viking landers during the mid-1970s, life on Mars was thought never to have existed. But, five years later Pathfinder *did* land on Mars in an ancient riverbed, and its cameras *did* find rocks that appeared to be sedimentary. And who knew what secrets those rocks might hold? The research and hypotheses of those two scientists took on new meaning as scenes of Mars were beamed back to Earth.

The two Mars researchers, Justin and P. J., were ordinary fifth-grade students who had stumbled upon some incredible ideas. They had confidence in their own ability to think scientifically, to analyze data, and predict possibilities. Justin and P. J. were indeed scientists.

Two

Getting Started

Getting inquiry started in the classroom is like building a campfire; it takes vigilance and perseverance. Making a campfire requires planning. Wood of various sizes must be gathered, kindling prepared, and materials assembled in such a way as to enable the small flame to survive and grow. Eventually, the campfire will take care of itself. Likewise, the goal with inquiry is for the children to eventually become independent as lifelong learners. But like the campfire, a fair amount of effort is required at the outset.

Surveys and Attitudes

At the start of a school year I have a classroom full of students of varying attitudes toward learning. To gather data about where my students are, I use an attitudinal survey. An example of one of these is shown in Figure 2-1. Attitudes of students are important. Knowing how kids feel can help fashion what we do in those early days of the year. Plus, the completed surveys can be used to measure changes later.

In addition to formal survey tools, dialogue journals can provide important information. As the children write about past experiences in science, the teacher becomes increasingly familiar with their interests, experiences, frustrations, and triumphs.

What Do Scientists Do?

Sometime during the first week of school one recent year, I asked the class, "What do scientists do?" Each table of five or six students had 180 seconds to silently pass around a paper on which to list things that scientists do. Some students saw this as a competition between tables, although it was not. As the seconds ticked by, I noticed

Student Survey

Name ______________________________

Read each statement and circle the appropriate response.

SA: strongly agree A: agree D: disagree SD: strongly disagree N: no opinion

	Statement					
1.	Learning is boring.	SA	A	D	SD	N
2.	I learn best by reading chapters and answering questions.	SA	A	D	SD	N
3.	As I learn, it is important to think about my thinking.	SA	A	D	SD	N
4.	I learn more if I have a choice about what I will be learning.	SA	A	D	SD	N
5.	When I talk things over with my partner I understand more about what I am learning.	SA	A	D	SD	N
6.	I learn more when I work in a group and share ideas.	SA	A	D	SD	N
7.	Discovering answers to my own questions is interesting.	SA	A	D	SD	N
8.	The best way to measure learning is for my teacher to give tests.	SA	A	D	SD	N
9.	My teacher can measure my learning by reading my journal.	SA	A	D	SD	N
10.	I like to discuss what I have discovered.	SA	A	D	SD	N
11.	Learning is finding out about things that interest me.	SA	A	D	SD	N
12.	Learning about science is only important for kids who want to become scientists.	SA	A	D	SD	N
13.	I am a scientist.	SA	A	D	SD	N
14.	I enjoy reading science picture books.	SA	A	D	SD	N
15.	A scientist asks questions.	SA	A	D	SD	N
16.	Science textbooks are the best books to read to learn about science.	SA	A	D	SD	N
17.	Scientists should answer old questions before asking new ones.	SA	A	D	SD	N
18.	I can learn more by reading than by doing.	SA	A	D	SD	N
19.	Facts I discover on my own are more memorable than facts someone tells me.	SA	A	D	SD	N
20.	Reading, math, and social studies are all parts of science.	SA	A	D	SD	N

21. What do you think science really is? (use the back for more space)

FIGURE 2-1 *Student Survey*

that some students passed the list along without adding to it while others could not wait for it to return to them so that they could add more.

When the time was up, the lists were collected, shuffled, and read to the entire class. "One group says that scientists do experiments," I began. "They also say scientists discover things, mix things together, ask questions, do research, read books, and write things down. A nice list," I added, "of things that scientists do."

I pulled another list from the pile and once again began reading each item to the class. " . . . Try to find things out, answer questions, make diagrams, list data . . . " Suddenly, I stopped.

"Wait a minute," I said. "I'm curious. Do any of *you* do these things? Let's start over and raise your hands if you do any of these things that scientists do."

I read back over the list. "Try to find things out." Many hands went up. "Ask questions." Many hands. Hands shot up for read books, make diagrams, investigate things, and more.

"Interesting," I said, pretending to be a bit confused. "I don't think you understood the directions. I didn't ask you to list things that *kids* do. The directions were to list things that *scientists* do!"

"But . . . " some of the kids began to say.

"Let's go on to another list," I interrupted. "Raise your hands if you do any of these things that scientists do. Work in groups, explore, make observations, do research, investigate." Hands were up all around the room.

"Clearly you didn't listen to the directions," I jokingly complained. By now a handful of students couldn't control themselves.

"But Mr. Pearce," one protested, "we do all those things, just like scientists do!"

"Yeah," spoke up more.

"Well, wait a minute," I said, acting as if I was processing a concept new to me. "Are you saying that all scientists are kids?"

"No!" came the uproar.

I then called on one little girl who was patiently waiting to speak with her hand up. "All scientists are not kids," she giggled. "But," she went on in a much more serious tone, "all kids are scientists, because we do the same things scientists do!" Heads were nodding all over the room. This class had made their first discovery of the new school year. Perhaps it would be one of their most important. Each and every one was trying to convince me of something that I, of course, had known all along: each kid was a scientist. They had been doing science all of their lives and were clearly ready to continue. How could I stand in their way?

At one time I thought that the ability to do science was a uniquely human trait. As humans, we are born scientists. We are curious, we explore, we ask questions, we investigate; we do all the things on those lists. But the notion that animals are not scientific might be questioned.

After the children convince me that they are all scientists, I like to read a true story told by Jean Craighead George about a crow named Crowbar her family adopted when the children were young. Crowbar actually played with the neighborhood children. One day Crowbar was trying to slide down the sliding board like the kids were doing. Crowbar found, however, that he could not slide. His feet in some way prevented him from going down the board. After looking about, the crow found a plastic coffee can lid in the sandbox, carried it over to the sliding board, and successfully rode it down. Might this be an example of an animal doing science? Perhaps. As the children consider whether or not Crowbar was a scientist, they are likewise considering the science in their own lives. If they ask, "Is Crowbar a scientist?," they are critically examining what it really means to do science and what it means to be a scientist.

The further our discussion goes, the more the children realize that indeed they really *are* scientists. "What do scientist do?" we ask. Many of the same things kids do.

The Question Board

Helping children to think of themselves as being scientifically literate means trusting them as scientists. Scientists ask questions; they do not simply answer questions in a book.

From the first day of the year, we have our question board posted someplace in our classroom. The question board is a laminated piece of poster board on which the children are invited to write, using overhead markers, any questions they may have. Class discussions, independent and assigned reading, and class activities can all elicit wonderful questions. When time is limited or I wish not to be the source of so many answers, or if I just don't know an answer, the student writes the question along with his or her name on the board. Often, during free moments of the day, a group of children will gather around the question board either adding questions or reading the questions of others.

Later, in three or four weeks, several student volunteers will copy the questions for typing. Each student in the class gets a printed copy of the questions. With that list we are able to analyze and discuss the questions. The question board becomes an integral part of the inquiry process throughout the year.

Questions: The Heart of Inquiry

In building that campfire, a match is the source of the flame. With inquiry science, questions spark the investigative process. Questions originate from many sources.

There are basically two types of questions: research (or read-to-find-out) questions and testable questions.

Research questions are the questions most often asked in school. They require students to use secondary sources of information such as books or computers. For example, a student could not readily answer the question "How hot is the surface of the Sun?" Sources other than the student would have to be relied upon to find an answer.

Testable questions are those questions students can answer on their own either through direct observation or by manipulating variables in an experimental setting. An example of a testable question would be, "Which evaporates faster, hot water or cold water?" A student could design an experiment and answer this question without referring to another source of information.

Of course, the differences between research and testable questions can sometimes blur. I remember a student who told me that every question was a testable question if the proper tools were available. I suppose he was correct. The important thing is to help children distinguish between research and testable questions based upon the practicality of finding solutions on their own.

Distinguishing between research and testable questions adds to our common language and builds a foundation for later investigations. The children enjoy looking over the question board and deciding which questions are testable and which are not. Those determined to be testable might be future topics of inquiry.

To show the children that they are already asking testable questions, we do the Question Search activity during the first or second week of school. Baskets of unusual objects are placed on the tables. Anything will do as long as the items are not familiar to the students. We have used odd-looking seashells, electrical components, seeds, interesting rocks, pieces of disassembled appliances, tools, and so on. Each student selects one item and examines it carefully. Then the activity sheet (Figure 2-2) is completed. First, the name, description, and a sketch of the item is recorded. Usually the name is unknown, so a made up name will do. Next, the student writes as many questions as he can about the object in the left column of the sheet. In the right column, possible sources of answers are listed.

A class discussion follows. Several students show their item and then read aloud some of their questions and sources. Sources mentioned may be books, experts, the teacher, a parent, or the computer. Someone always mentions *doing* something to find an answer to a question. Those questions, of course, would be the testable questions. Together, the class can decide not only which questions are testable, but what kinds of experiments or investigations could be designed to answer those questions.

Younger students, when asked to think about their own testable questions, often begin with "Can I . . . ?" type questions. "Can I blow big bubbles?" or "Can I build a tall tower?" These are great starters, but we do want to eventually guide our students into asking and phrasing their questions in ways that will lead to more meaningful investigations. An initial response to the questions above would be for the teacher to ask, "*How* big a bubble?" or "*How* tall a tower?" Encouraging children to quantify at an early stage will greatly help them later.

The phrasing of the question can mean a lot as well. Rather than asking a *yes* or *no* question about blowing a big bubble, the student could be encouraged to ask a similar question: "When *comparing* bubble solution A to bubble solution B, which will make the biggest bubbles?" Here we are helping the student to think about variables in an experimental design. As a minilesson to help students refine their testable questions, I use the More Testable Questions worksheet (Figure 2-3). Together we

Question Search

Name ____

Select an object. Write its name (or make up a name). __________________

Write a *description* of your object.

____

____

____

____

____

____

Make a *sketch* of your object. Use labels to help describe the object.

Write as many *questions* about your object that you can think of.	List ways you might be able to find *answers* for your questions.

FIGURE 2-2 *Question Search*

More Testable Questions

Name ______________________________________

We have been asking and writing ______________ questions. These are questions that you can answer yourself by _________ something. Sometimes the way your questions are worded will lead to a more interesting investigation.

Instead of *"Can I . . . ?"* questions, try *"Is it possible to . . . ?"* questions.

Is it possible to clean polluted water?
Is it possible to make glue that will stick under water?
Is it possible to grow plants using salt water?
Try one of your own.____

Try using *comparing* questions.

When comparing ________ *with* _________, *which will* ________?
When comparing sunflowers with suet, *which will* attract the most birds?
When comparing C batteries with D batteries, *which will* light a bulb brighter? or longer?
When comparing radish seeds with grass seeds, *which will* sprout sooner?
Try one of your own. ____

Use *"What if . . . ?"* questions.

What if I plant a seed in a flower pot and put the pot into a dark closet?
What if we put vinegar and baking soda into a balloon?
What if we poured oil into a model ocean?
Try one of your own. ____

Try using *"How can we . . . ?"* questions.

How can we prevent erosion on our model mountain?
How can we build a boat that will hold 300 grams?
How can we maintain a healthy terrarium?
Try one of your own. ____

Another kind of question to try is the *"What is . . . ?"* question.

What is the life span of a mealworm?
What are the stages of a monarch butterfly's development?
What is the most effective way to blow bubbles?
Try one of your own. ____

Other *INQUIRY* question starters are:

If I had _________, *how could I* _________________?
How can I improve ____________________________?
What will happen if ___________________________?
Suppose I could ______________________________?

FIGURE 2-3 *More Testable Questions*

examine different ways to write testable questions and model for one another possible questions to ask.

Early in the year is the time to find out about the scientists in our room. In addition to class discussions and dialogue journal entries, students might complete a survey similar to Science and Questions, as shown in Figure 2-4. This survey gives students an opportunity to tell about past science topics studied and to list questions that those units may have brought up but provided no time to explore. It also serves as a means for students to examine what they would like to know or try out. Tapping the past is crucial in order for inquiry to grow.

Early Readings

As important as testable questions may be in leading us to discoveries through our own investigations, secondary sources of information are vital. Before any questions can be asked, some degree of basic knowledge must be present. How can a student ask questions about seeds if she has no idea what seeds are?

An assortment of books in the classroom can provide the prior knowledge needed to launch young scientists on their investigations. Those old science textbooks will be rarely visited when students are given a choice. It's the trade books that attract and interest. Short science books with many pictures are ideal sources of information. Even though many may be labeled for younger ages, older students will find in them valuable information. We call our science trade books "top-shelf" books.

Our top shelf contains books by Seymour Simon, Paul Showers, Aliki, Franklyn Branley, and others. They are all of high interest and most contain many pictures and diagrams. Some are borrowed from the school library or are shared with other teachers in a rotation.

We use our top-shelf books for independent reading, student-generated book reviews, book talks by our library committee, or as part of a more traditional assigned-reading program. My goal is for the children to become familiar with the books, enjoy reading them, and ultimately think of testable questions as a result of them. After reading *Germs Make Me Sick* by Melvin Berger, a student wanted to know "Which germs make me sick?" After reading Franklyn Branley's *The Planets in Our Solar System*, another student asked, "How many times can I see a planet in one week?"

During those first few weeks, I like to use some of the top-shelf books for read-aloud (see the list of resources for some favorite read-aloud books). As I read to the class, I stop and think aloud, modeling the ways readers process what is read. I also verbalize questions that come to mind as I read. I might even go over to the question board and write down a question or two as the class watches.

One day early in the year, I was reading *A Snake-Lover's Diary* by Barbara Brenner. The story is about a boy who convinces his parents to let him keep a snake he has captured. His little brother, thinking the snake lonely in its new cage, puts in his

Science and Questions

Name ____

Think of some science topics you studied in school earlier this year or in past years. Complete the spaces below.

1. One science topic we studied was ________________________.
   I am still wondering (write two or three testable questions):

2. Another topic we studied was ________________________. Write two or three "what if" questions:
   What if. . .

   What if . . .

   What if . . .

3. Another science topic I remember studying was ______________________.
   I never had a chance to try (write two or three things you would like to do):

4. I remember another science topic was _________________________.
   (Think about the materials you used. If you could have any of those materials again, what would they be and what would you do with them?)
   I would like to have ____ materials.
   Then I would:

Use the back if you need more space.

FIGURE 2-4 *Science and Questions*

Kids' Inquiry Conference Journal Article

Evaluation

Name ______________________________

An effective KIC article contains several sections. Each section tells the reader something about the question, investigation, or discovery made by the author.

Read a KIC article and answer the questions in the spaces below.
Article Title ______________________________
Author ____________________ Year ________

1. Did the author describe the TOPIC of the investigation? YES NO
   What was the TOPIC of the article you read? ____________________
   ____
   Did the author explain WHY the topic was chosen? YES NO
2. Did the author write the exact QUESTION that was trying to be answered? YES NO
   What was the QUESTION being answered? ____________________
   ____
3. Did the author include some BACKGROUND INFORMATION about the topic? (*Background information* is information about the topic, found in books or other articles, which tells more about the topic being investigated.) YES NO
4. Did the author explain *exactly* what was done to ANSWER the question? (The author should describe the steps followed during the investigation so clearly that the reader could try the same experiment.) YES NO
   What did the author do to answer the question? First, ____________________
   ____
   Next, ______________________________
   ____
   Then, ______________________________
   ____
5. What FACTS did the author discover during the investigation? (Try to list at least three.)
   a. ______________________________
   b. ______________________________
   c. ______________________________
6. Did the author suggest some ideas a future student might try NEXT YEAR? YES NO
   On the back, write what the author suggested for future students to try.
   ____

Based upon your answers, evaluate how well the author wrote the article you read. Circle a grade: A B C D E Then, explain your grade on the back.

Thank you for your answers!

FIGURE 2-5 *Kids' Inquiry Conference Journal Article Evaluations*

own pet frog to keep the snake company. Of course, the frog disappears. I wondered aloud about snakes and frogs and what they might eat. "Do snakes eat frogs?" I asked. "If so, how do snakes capture frogs? How fast are frogs? How far can they jump?" Someone suggested I write my questions on the question board, which I immediately did. All of this encourages students to read, think, and to develop and record questions, taking steps along the path toward inquiry.

When kids engage in investigations based upon their own questions, they have stories to tell and discoveries to share. Inquiry science provides authentic reasons to read and to write. Early in the year we make use of the Kids' Inquiry Conference journals, articles written by earlier students who presented their investigations and discoveries at the Kids' Inquiry Conference (see Chapter 9 for a description of KIC). Early in the year, the KIC journals provide models of documentation and examples of processes that other students used to look for answers to their questions. Then, as critical readers, current students use the KIC article evaluation form (Figure 2-5) to assess one of the articles read. Using the form provides an opportunity for the children to consider the characteristics of a quality article. These kinds of activities will help the children when they become KIC authors themselves later in the year. The use of KIC journals as instructional tools demonstrates to the students that I value them as scientists and as authors.

These early activities are important in establishing the value of student questions and the inquiry process in our classroom. But for that flame to really start to take hold, we have to get our hands dirty. To do so we undertake our first inquiry period.

Inquiry in Action: First-Day Discoveries

The first day of a new school year is not too early to begin inquiry. Something simple yet open-ended is ideal to hook the students on the excitement of science.

One recent year I instructed the children to get four white beads from a bowl on the table and to cut a six-inch length of string. Scissors and rulers were available. (I learned a lot about the class's ability level just by seeing who was able to measure and cut accurately.) The children were to place the beads on the string, tie the two ends, and observe their beads throughout the day.

"Is that all?" one student asked.

"Just watch closely," I repeated.

The day went on with all the usual first-day activities. Many of the children forgot about their beads. Suddenly Joe, who was seated by the windows, excitedly raised his hand to tell the class that by shaking his beads, he was able to make them change color. He continued shaking the beads back and forth on the string. Sure enough, the white beads had changed to purple and pink and yellow! Everyone else tried to make their beads do the same. Beads were clicking all over the room.

"You're right," said someone near Joe, "shaking them *does* make the beads change." Everyone around Joe was nodding their heads and congratulating him on his discovery.

"But mine don't work!" complained a student from across the room, who was shaking her beads as hard as she could. "Look, mine are still white."

After a lot of shaking, some of the students were excited about their beads' transformation while others were disappointed, having not gotten similar results. I told them to be patient, that all the beads were the same.

The beads were eventually put away and we went on to other things. It wasn't until recess after lunch that the truth about the beads was discovered. These were no ordinary beads, and they certainly did not change colors as a result of being shaken. Outside, the children made some interesting discoveries. In their closed hands, the beads were white, but exposed to sunlight, they changed to different colors. Plus, their experiments with shaking and not shaking in the sun and in the shade indicated that movement had no affect on the beads. It was the sun, they told me, that was producing the change. Next, all the kids wanted to go back inside to see if the lights in the classroom would have the same effect.

I had given the children UV-sensitive beads, which change color in the presence of UV (ulraviolet) rays. When the sunlight had come through the window near his desk, where he was shaking his beads, Joe had made an interesting discovery, which many students had accepted. However, when more data were gathered later, the students learned that what had seemed reasonable before was no longer valid. An important lesson learned on the first day, not from their teacher, but from one another.

The endless stream of questions and experiments flowed for several weeks. Some children experimented in their swimming pools to see how deep UV rays travel in water. Others tried filtering the sunlight through different sunglasses, and even put sunscreen of varying strengths on the beads. My initial goal was to get them to ask questions. They did that and more!

(See the list of references and resources at the end of this book for inexpensive sources of UV-detecting beads.)

Three

Our First Inquiry Period

At this early point in the year, I have established with the kids that each is a scientist. Since they are scientists, their questions are valued and seen as avenues toward possible investigation. The overarching goal as we proceed through the process of initiating inquiry is to help the children *(a)* ask testable questions, *(b)* develop investigations to answer those questions, *(c)* make discoveries through their investigations, and *(d)* share their findings with their peers. Our first inquiry period helps address each of these goals.

To plan for inquiry (and careful planning is essential), it is important to think about meeting each student where he or she may be. Early in the year, I assess the individual strengths and needs of our students so that the activities can be designed to provide flexible opportunities to reach each child in the classroom. I know that some students will enthusiastically participate in spite of what I do, while the more reluctant or inhibited will do so only *because* of what I provide. Part of the early assessment occurs during the inquiry period itself, when I watch what the children are doing and saying, and when I am able to observe where their questions and curiosities take them in their explorations.

The initial inquiry period can help with assessing how to address the needs of each student. It consists of several hands-on science activities located around the room. Each is designed to elicit lots of questions, tap prior knowledge, and intrigue students with both new and more familiar topics and activities. This inquiry period is an early start toward a more independent student-directed inquiry process to follow later.

At the beginning of the inquiry period, I introduce each area to the class. The students are encouraged to write down in their dialogue journals any testable questions that they may think of based upon the topics described. These initial questions serve as starting points for inquiry. A child may not answer that starting question, but no one asks, "What should I do?" because each has a framework for discovery. These first questions may be as simple as, "How tall can I build a tower?" but they are the same kinds of questions that drive scientific exploration everywhere.

The choices at the stations vary but may include such topics as magnets, ramps and rollers, boatbuilding, bubbles, microscopes, structures, and more. Some may be curriculum related from previous years; others may be new. Included at each location are a variety of open-ended possibilities along with more teacher-directed challenges. I am building a scaffold that will help make the transition from a more traditional approach seen here in our early inquiry periods toward those more independent investigations to follow.

Described here are some of the activities, along with sample cover sheets, which are posted at each location. A look through curriculum guides from previous grade levels can provide additional ideas. The materials needed may be hidden away in nearby closets. (See the list of references and resources at the end of this book for more ideas.)

Magnets

This area contains an assortment of magnets: doughnut, bar, horseshoe, and any others available. With the magnets are a variety of magnetic and nonmagnetic items: paper clips, nuts, bolts, coins, staples, keys, and so on. I also include one or two little toy compasses that are affected by the magnets. The idea is for the children not only to use the magnets in ways learned earlier, but also to try using the materials in ways they may have wondered about during their past experiences with magnets. A cover sheet is posted with some ideas and possible questions (see Figure 3-1).

The suggested ideas and questions are simple. They are designed for students who may not have much experience with magnets, yet are broad enough to challenge the students who do.

For this activity there is nothing for the students to write. At this early stage in our progression of inquiry, I want the children to be engaged and intrigued. The recording on paper has its place later.

Boatbuilding

Boatbuilding goes way back to early childhood experiences in the bathtub. Sink/float units in the primary grades never seem quite long enough, and experimentation with buoyancy is a treat for all ages. This activity provides a wide range of possibilities. Students are invited to use materials to design and build original boats and then test their boats with cargo. Unrestricted explorations permit the kids to learn about the materials and how they can be fashioned to float. Challenges make use of what the children already know, and are similar to contests with rules and specific guidelines. Materials for boatbuilding include tubs (new kitty litter pans work well), clay,

Magnets

Examine the different magnets.
Which magnet do you think is the most powerful?
How can you find out?

Try picking up paper clips with a magnet.
How many clips will *one* magnet pick up?

Place two magnets together.
How many clips do you think the two magnets will pick up?

Try it!

How about three?

Place two or more doughnut magnets on a pencil.
What do you observe?

Try some other ideas of your own.

FIGURE 3-1 *Magnets*

aluminum foil, straws, Popsicle sticks, rubber bands, sponges cut into small blocks, and cargo (fish weights, coins, paper clips, marbles, etc.). Cover sheets might be similar to those shown in Figures 3-2 and 3-3. While the sink/float activity may seem like a primary time filler, in the hands of student scientists, it becomes a venue for sophisticated inquiry. This year, Earl chose to wrap a sponge in aluminum foil. "I think," he said, "that this way the sponge will displace the water without absorbing it. Then my boat might hold more cargo."

The students are then invited to use what they learned (or already knew) to enter a boatbuilding challenge. A sample cover sheet for a challenge is shown in Figure 3-3. A sample Hall of Fame is shown in Figure 3-4.

Other challenges include the clay boat challenge, in which students make vessels using twenty-five grams of clay and then test their boats with varying amounts of cargo. We have also used the aluminum foil challenge in which a six-inch square of foil is the only material permitted. Blending controlled challenges with broader explorations provides different kinds of opportunities for the students. A student might immediately use something learned from one kind of activity when doing another. The chain of investigation, discovery, and technology starts to come together here!

Boatbuilding

1. Design a boat using aluminum foil, clay, or other materials.

2. Test your boat to see how well it floats.

3. Choose a cargo to load on your boat. You might use fish weights, marbles, pennies, or something else.

4. Think about how and where to load the cargo on your boat. Then, have a test voyage.

5. Redesign and retest your boat to increase its capacity.

6. Next, you might want to use what you have learned to enter a challenge!

Good Luck!

FIGURE 3-2 *Boatbuilding*

Boatbuilding Challenge

1. Using no more than 6 straws, 6 Popsicle sticks, and 6 rubber bands, construct a boat.

2. See how much cargo your boat can carry. Weigh your cargo.

3. Place your name on the

BOATBUILDING

HALL OF FAME!

Include the weight of your cargo!

FIGURE 3-3 *Boatbuilding Challenge*

Boatbuilding

Hall of Fame
(6 straws, 6 Popsicle sticks, 6 rubber bands)
Weigh (or count) the cargo carried by your boat.
Enter your name below!

NAME	*DATE*	*CARGO WEIGHT (or count)*

Congratulations!

FIGURE 3-4 *Boatbuilding Hall of Fame*

The Hall of Fame charts (a sample is shown in Figure 3-4) are one way for student documentation of achievements with this and other activities. Hall of Fame criteria should be relatively easy so that many students get to sign their names. Some invite all participants to sign in so they can document what was accomplished. Then, as other students visit the activity, they can see what and how others have done before them. Rather than a competition, the Hall of Fame really serves as a means of student communication and bonding. These concepts become extremely important later as we consider our growing community of scientists.

Ramps and Rollers

This area is about as open-ended as an activity can be. Students are invited to construct ramps upon which marbles, Superballs, Ping-Pong balls, and other round objects are released to travel a route to the floor. For the ramps, we use cardboard toilet tissue, paper towel, and gift wrap tubes cut in half lengthwise. Then, with tacks and masking tape, the tubes are attached to a wall to make ramps of different designs.

The challenge here involves using a stopwatch to time the travel of the ball from point of release to the floor. The longer the fall, the better. The students enjoy changing their designs to lengthen the transit time. A very clear cause-and-effect relationship can be viewed. An interesting degree of engineering, troubleshooting, and problem solving can be observed.

Other Possibilities

Structures offer another wonderful topic for exploration. We have constructed forms using toothpicks and miniature marshmallows—old, familiar materials. We have also experimented with toothpicks and beans. (The beans are purchased dry in bags and are soaked overnight. As they dry following construction, the beans contract around each toothpick, which adds stability to the frame.) Straws and paper clips work well, too. Paper clips bent at ninety-degree angles can be wedged into the ends of straws.

These materials have been used to build towers, bridges, and buildings. Challenges consist of supporting one- and three-pound weights made of small coffee tins filled with sand. Students can also try to support a one-pound weight three inches above the tabletop using toothpicks and beans, or toothpicks and marshmallows. Straws and paper clips are used to build a structure that holds a three-pound weight five inches above the table. Attempts usually turn into group efforts, with those successful teams being added to the various Halls of Fame.

If a digital scale is available, it makes a really useful station. We use a relatively inexpensive model. Using the scale, students have discovered that coins vary in weight, that old markers weigh less than fresh ones, and that hot water can be measured evaporating from soaked paper towels. Students from boatbuilding also make use of the scale to measure the weight of their cargo. Our scale, which measures to a hundredth of a gram, can also be used to measure the weights of insects, leaves, and seeds. The questions that come from these kinds of activities are endless.

Bubbles have been a longtime favorite for inquiry activities. Questions arise concerning ingredients (soap brands and proportions, the use of glycerin, vinegar, sugar, etc.). Different tools offer a variety of ways of making bubbles. Students have wondered about bubble sizes, bubble life spans, distances bubbles travel, and more. For this initial inquiry period, we have offered two bubble solutions for students to test and compare. Numerous resources are available which discuss solutions and bubble-making tools.

A look through curriculum guides from previous grade levels can provide additional ideas. The materials needed for inquiry may be hidden away in nearby closets. See the list of references and resources at the end of this book for more ideas.

During the inquiry period, the teacher circulates from group to group. I enjoy simply listening to the conversations. Much can be learned about each child through these opportunities to ponder, question, and reflect. Are they able to explain what they are doing? Do they have a plan? What is intriguing them? I also use these occasions to chat with students working alone. Some may feel excluded from a group and may want to share their activity with someone. Others working alone may do so because they prefer the focus and intensity of a self-directed project. As a visiting observer, the teacher's role is to be available to listen, to assist when asked, and to locate any additional materials that might be needed.

Logistics

In our class, we conduct several of these structured-activity inquiry periods before moving on to further stages. For these periods I generally do not assign or have students sign up for particular activities. Having the children move about and form their own flexible groups at each location has worked well. Of course, the dynamics of some class groups may require adjustments for control and order.

Some teachers take a slightly different approach, with one inquiry period undertaken per week, each with its own particular topic. For example, one Wednesday afternoon might feature magnets and another afternoon might center around sink/float. This approach does not require a variety of materials to be gathered at one time, but does necessitate a supply of more of the particular materials being explored that week. The age and temperament of the students will help to decide which method might be most effective. Keep in mind that there really are no right or wrong approaches. Providing children with provocative materials and opportunities for questioning and investigating are the only true tests of the merit of any inquiry plan.

Materials

Materials for inquiry can come from many different sources. Curriculum-related materials will often be found in a school storeroom. It is surprising the amount of supplies going unused (and often unknown), hidden away in back rooms and forgotten closets. Things are ordered by teachers or administrators who then leave the school, and those items are put in storage. Many teachers find useful items in the trash.

Middle or high schools will often send extra or discarded materials to the local elementary school. Parents, too, can be a source of items to use. A letter home asking for a variety of common materials will often yield a rich return.

Another source of materials is local (or even distant) businesses and corporations. Some have unadvertised policies of giving teachers whatever items they can provide that might otherwise be discarded or placed into permanent storage. Some businesses will even offer nonsurplus items to teachers or schools. One international corporation we know of has a policy of directing its personnel to respond in any way possible to all teacher requests.

Local hospitals can likewise provide a variety of items. Because of expiration dates or broken seals, things like tubing, markers, plastic pipettes, calibrated medicine cups, and other items must be discarded. Of course, these materials are unused and good as new for classroom use. With so many stringent regulations in a variety of industries, perfectly good, unused products are available free of charge. Scouting out local businesses would be an excellent project for parent volunteers, leaving the classroom teacher more time to plan the next inquiry period.

Student Documentation

As scientists, students are expected to not only ask questions, investigate, and explore, but also to document. Early in the inquiry process, documentation demands are limited so as not to interfere with any investigations under way. Some form of sharing and writing down observations and discoveries, however, really does need to occur to set a tone for the seriousness of our endeavors. Inquiry may appear to be (and often feels like) play, but it is not.

The children wrote down several testable questions at the start of our inquiry period. It is interesting to find out if those questions were pursued and answered. After the cleanup has been completed (a sometimes arduous task), there comes a time for calm and quiet as everyone sits down to reflect. They all have stories to tell about what they had done and are eager to talk about their investigations.

"I wrote down a question," said Justin, "about boats. I wanted to know if a boat with steeper sides could carry more cargo than one that had low sides."

"What did you find out?" I asked.

"Well, a boat with steep sides can carry more cargo, but when it is empty it tips over easier."

Someone asked what the student had done to find this out. Justin went on to explain (and show) the construction of his aluminum foil boat. The vessel held over three hundred grams of weight before it sunk. Other designs with lower sides held less.

Heather told the class about her experiences with toothpicks and marshmallows. "I heard that triangles work better than squares," she reported. "I saw some others having a hard time with squares, so I tried triangles and they seemed to work."

"What could you try another time?" I asked.

"I would like to see if triangles work better with other materials, like the beans or straws," she said.

It is always exciting when the students extend the activities provided to include their own thinking and their own imaginations. Watching the flickering flame take hold and grow is one of the teacher's rewards for all the effort.

Four

Discovery Boxes

Another component of inquiry science in our classroom is the discovery box. The differences between the early inquiry periods and the use of discovery boxes are subtle. For the children, discovery boxes offer a wider range of topics. One student mentioned that early in the year we *played* science; now we were really *doing* science.

Whatever the distinction, discovery boxes continue the spiral of inquiry. They integrate more reading and documenting while providing more choices.

The concept of discovery boxes arose the summer I attended the Elementary Science Integration Project. ESIP was a project, sponsored by the National Science Foundation, in which concepts of curricular integration and inquiry in the classroom were being examined by teachers through a variety of experiences. One day we divided up into small groups. Our group was assigned a sink/float kit filled with all kinds of materials. We were to test the activities in the kit to see if they would work well for second- and third-grade students. Later, we were to report back to the larger group with the results of our experiences. In addition to containing a lot of interesting materials, the kit also had a packet of cards full of step-by-step directions. Our mission: follow the directions and keep notes about our progress.

One hour later, we were called to report to the larger group. The time had flown by, and we were embarrassed to say that we had not completed even the first activity. All we did, we told the group, was take the materials out of the box and use them in some sink-and-float experiments of our own. The questions we asked each other seemed far more intriguing than any on the cards. We had not accomplished our assigned task, but along the way we had made some interesting discoveries.

Later, our group realized that what we had done during that hour may have been seen by some colleagues as being off task. Yet, what we had *accomplished* in that hour went beyond merely following directions. The questions we were asking and investigating were our own. Likewise, the discoveries we were making belonged to us. We soon realized that the real discovery we had made that afternoon was that science isn't necessarily about following directions.

What if we tried giving students similar boxes of materials with no directions to follow, no packets of activity cards? How could we help students develop *their own* testable questions? Would students be able to devise their own investigations based upon their own questions? How could we encourage students to document their discoveries for others to see? Would the children find the experience as exhilarating and exciting as we had? And how could we ever assess student progress as they worked through this process? These were tough questions to ponder, but the possibilities that came from that afternoon soon led to the development of discovery boxes in my classroom.

When I first conceived the idea, discovery boxes seemed to be a natural extension of hands-on science activities. These boxes appeared to be a new destination. Now, however, the boxes, like hands-on science itself, are merely part of the inquiry spiral that take the children to higher levels of inquiry thinking and discovery. The discovery boxes have become an important transition from early inquiry periods at the start of the year to more sophisticated ongoing investigations that follow later.

What Is a Discovery Box?

"Mr. Pearce! Mr. Pearce! You have to see this! I found another rock that is magnetic!"

"Mr. Pearce, I need a container for the glue I invented."

"Mr. Pearce, will the light bulb glow brighter if I use three batteries?"

Inside a discovery box are worlds to discover. A typical discovery box contains materials based upon a particular theme or topic. The materials might be assembled with certain activities in mind, but ideally they are put together with the intention of allowing the children to devise their own use of the contents. Also included are a variety of trade books on the topic and a folder that contains a cover sheet and log pages for documentation. *Not* included are directions for specific activities.

For example, an electricity discovery box might contain wires, light bulbs, batteries, battery holders, small switches, motors, buzzers, and more. Common items like aluminum foil, paper clips, and balloons could also be included. If the children have previously studied electricity, they should have the prior knowledge to make use of many of the items in the box. The elements could come from the electricity materials used for that unit of study. An electricity box could be used even if the children did not study electricity earlier. Some might have experience from home, other schools, or elsewhere.

But every child has not had prior experience with all the materials. Here is another opportunity to inspire children to read for information. Trade books are the key. The electricity discovery box would contain several trade books on electricity. These serve to refresh the memories of children who worked with electricity before, or to inspire new questions with ideas and suggestions. The books provide the literature link so important to inquiry science. As the children explore and investigate, they are drawn to the books for authentic reasons.

Finally, the electricity box contains a folder with several forms for documentation. The first form is a sign-in sheet for students to use when they begin using the box. A cover sheet (see Figure 4-1) tells the students what the box may contain and encourages the students to think of their own questions. Several questions are suggested for students having difficulty thinking of their own. A log sheet (Figure 4-3) is provided for students to document their question, investigation, and discoveries. Proper completion of the log sheets will enable later students to read what was done and then possibly replicate or build upon those activities.

Other topics could be taken from the curriculum. It is important to keep in mind that once a unit has been taught, its materials need not be immediately put away. The students need the time to investigate, especially with those familiar topics for which there were so many questions and so little time to explore!

How Are Discovery Boxes Used?

Before the discovery boxes are introduced, some preliminary work must be done. Figure 4-2 shows a survey that is completed by the students to help them think about previous science topics with which they may have worked. By asking testable questions on the survey, the students begin thinking about using the boxes for their own investigations.

Several days before the discovery boxes are used, the children sign up for their preferences. The limited number of spaces for each box keeps the groups small. Then, after the boxes have been selected, the children read through the trade books and folders to get acquainted again with the chosen topic. DEAR (drop everything and read) time a day or two before is ideal. The room is quiet, the children are interested in their selected topic, and they are eager to read the related books. Browsing through the folders and seeing what other students have done in the past is interesting for them. Often, the children will recognize names of students from previous years. They always seem interested in what other children have done. The process of reading the books and folders before the boxes are used helps to plant more ideas about some possible questions to pursue when it is time to work with the boxes.

Before we use the discovery boxes, the children make entries in their dialogue journals. The entries include the date, the box they will be using, and one or more questions they hope to answer. This page in their journals can also be used to record any notes during or after the inquiry period. Later, the students will write about their experiences. Other discovery boxes that have worked well are detailed below.

Science Discovery
Electricity Discovery Box

Some materials you may find in this discovery box:

- batteries (various sizes and ages)
- wire
- aluminum foil
- motors
- buzzer
- bulbs and holders
- switches
- paper clips
- balloons
- wool fabric

Directions

1. Consider a question that you would like to answer. You may select a question from below or you may develop your own.
2. Read some of the SCIENCE DISCOVERY LOG forms in this folder to see what other student scientists have investigated.
3. Begin a SCIENCE DISCOVERY LOG sheet of your own.
4. Conduct your investigation.
5. Complete the SCIENCE DISCOVERY LOG sheet that you started.
6. Add what you discovered to the Book of Discoveries.

Some questions that you may want to investigate:

- Which batteries make the light bulbs light the brightest?
- Which batteries last the longest?
- How can circuits be wired in different ways?
- What if we hook many bulbs to one battery?
- What if we hook many batteries to one bulb?
- How many bulbs will one battery light?
- Will electricity pass through water, salt water, colored water?

Remember, the best questions are your own questions!
Be curious! Be creative! Have fun!

FIGURE 4-1 *A Sample Discovery Box Cover Sheet*

Liquids and Solids

This box provides students with opportunities to mix water, food coloring, sugar, and salt. Students also explore acids using vinegar and pH paper. Coffee filters, strainers, litmus paper, thermometers, baking soda, stirrers, eyedroppers, flour, and paper cups of various sizes are also included. A cover sheet similar to the electricity sheet is prepared for the folder.

Possible questions might include:

- What if a drop of food coloring is placed in hot water and not stirred?
- What if a drop of food coloring is placed in cold water and not stirred?
- In what temperature water does sugar or salt dissolve the fastest?
- What if we mix baking soda with vinegar?
- How can we use litmus paper to measure the strength of vinegar as an acid?
- Can colored water be filtered to make it clear?
- Can dirty water be filtered to make it clean?

The best questions, however, are those elicited from the students.

Mealworms

These little animals are the perfect critters to investigate in the classroom. They are tough, harmless, and exhibit the four stages of development (egg, larva, pupa, and adult). Mealworms are inexpensive and easily obtained from most pet stores. They can be stored in a container of dry oatmeal (with an occasional tiny sliver of apple for moisture) for a long time. Properly cared for, a mealworm colony will produce several generations in a school year.

Possible questions to ask about mealworms might include:

- Do mealworms prefer light or dark?
- What foods do mealworms seem to prefer?
- Are mealworms curious?
- How long is each stage of a mealworm's development?
- Can mealworms be trained?
- What do mealworms do when they encounter an obstacle?

Other topics for discovery boxes:

Air and Flight
Art
Concoctions
Hourglass Workshop
Light and Color
Mineral Testing
Mountain (Erosion)
River
Structures
Take-Apart Box
Tower Workshop

Inquiry time with discovery boxes usually occurs once every other week and consists of a fifty-to-sixty-minute period. Having already thought about questions,

Introduction to Discovery Boxes

Name ________________________________

Soon our class will be using DISCOVERY BOXES. You will have opportunities to ask testable questions and investigate answers. To begin, look at the list of topics in the column below. You may have studied many of them in science in past years. After looking over the list, follow the directions.

Select any ____ topics. Write down one or more testable questions you would like to investigate.

ELECTRICITY ________________________________

MEALWORMS ________________________________

FLIGHT ________________________________

STRUCTURES ________________________________

LIGHT AND COLOR ________________________________

LIQUIDS AND SOLIDS ________________________________

CHEMISTRY ________________________________

MAGNETS ________________________________

HOURGLASS WORKSHOP ________________________________

BUBBLES ________________________________

BOATBUILDING ________________________________

In a few days we will be using the boxes for our own investigations. Before then you will be asked to look through some books on your chosen topic. Think about some other questions you would like to explore!

FIGURE 4-2 *Introduction to Discovery Boxes*

perused the trade books, and read what other students had accomplished, the children are ready to begin their own investigations.

When each small group receives their box, they are reminded to sign in on the first page in the folder. This not only helps me later with assessment (tracking who has used which box), but if the box is damaged by careless use, I know who is responsible.

The students are required to work with their group only on the box for which they signed up. This helps to maintain focus and minimize intrusion by others.

The students also complete the log sheet in the folder (see Figure 4-3). With the freedom of autonomy comes the responsibility of documentation. Their initial question is recorded and then a description (with a sketch and labels) tells about the investigation and discoveries. Each student may complete his or her own log page or work collaboratively with partners. I do make them aware that the pages will be assessed. (See Chapter 10 for more about assessment.)

Occasionally a group may run out of ideas before the time is over. After they have completed their log page(s) and have cleaned up their materials, they may visit other groups, but only as observers. Often the observers will ask questions that lead to interesting discussions. Also, as observers see other boxes being used, they will have a better idea about which to sign up for next time.

The excitement during these discovery box periods is incredible. The classroom teacher has probably never heard her name called so often in such a short period of time. All around the room, discoveries are being made, and everyone wants to share their observations with their teacher.

Traveling from group to group as a visiting observer may be a new role for some teachers. Instead of checking to see if the children are following directions and doing it "right," the teacher steps back and listens in on what is being investigated. Teachers need to allow students the luxury of time to revise their thinking so that they–not the teacher—own the experiences. This is exciting for the kids, who are eager to share what they are discovering.

Owning discoveries means sometimes starting off in the wrong direction. For example, while working with the Rocks and Minerals box, Danielle came up with an exciting discovery: her piece of mica was magnetic. She came dashing across the room, mica in hand, to show me that her mineral stuck to a magnet.

"I found out that if you put vinegar on mica, it becomes magnetic," she said. "Look, see—it's magnetic."

I didn't want her to keep this misconception, but I didn't want to step in and tell her that she was wrong, so I asked a question. "What if you try another flat surface, other than a magnet? Will the mica stick to it, too?" Ten minutes later, Danielle came back with an updated discovery. "The vinegar didn't make it magnetic, Mr. Pearce. It stuck because it was wet. I tried it on a couple of things. See, it sticks to the bottom of this plastic cup." Danielle's discoveries were both interesting, but it was her process that was important. With time, materials, and questions, children can revise their misconceptions just as adult scientists have done for ages.

Science Discovery Log

Activity ______________________ Names ________________ ________________

Date __________ ________________ ________________

What question did you try to answer?

Explain what you did to answer your question.

Make a sketch of your experiment.

What did you discover today?

What new question are you curious about for another time?

Are you pleased with your results today? YES ___ NO ___ NOT SURE ___

How would your group rate this activity? Great 10 9 8 7 6 5 4 3 2 1 0 Terrible

FIGURE 4-3 *Discovery Box Log Page*

The time soon comes for cleanup. For many it is far too soon. If a definite time limit is established, cleanup becomes almost a game, with everyone racing the clock. Of course, the kids must be reminded that speed does not make up for a sloppy job. This is where a plan for box management is essential! If the students know that each box will be checked, they will be careful to do their best. Having a small committee of students whose job it is to put the straightened boxes back on the shelf (and inspect each box as they do so) frees the teacher to patrol the room and offer assistance where needed. When cleanup is complete, those students on the committee can call students back who returned messy boxes, and can later give the teacher a list of consumables that must be replaced.

Finally, everyone is seated, usually at the desks of others, since each group stays together for class debriefing. The children have the folders from the boxes and any products from their investigations. Each group has a minute or so to tell the class about their experiences. This is an important component of the discovery box period. Not only must the students conceptualize and verbalize their own ideas and perceptions, but those listening always compare their own experiences with what is being described. The teacher can use this time to correct through questioning improper testing methods, lack of controls, or serious misconceptions. I always ask, "How do you know this? What comes next? Where do you plan to go with your discoveries?" Furthermore, as each group tells about their investigation, the discovery box is advertised to the class to help with box selection next time. The discussion, questions from the class, and ensuing dialogue can be quite rich. Those speaking are experts; those listening want to hear more.

Sometimes, two groups are investigating similar questions. During debriefing one day, Shannon described her attempt to make stamps using vinegar and Elmer's glue. This mixture, in the right proportions, gives ordinary paper the quality of postal stamps. After hearing her story, members of another group whose goal was to make glue with materials in the liquids box, became very interested. They asked questions about Shannon's success and took notes on her comments.

Inquiry period with the discovery boxes ends with the students adding to their dialogue journals. Reflecting upon their investigations, the children write about their observations and adventures. Was the original question answered? Were there any surprising observations? What was discovered? Where might they go next with this topic of investigation?

Journal entries are important sources of information. I treasure the quiet time after the controlled chaos of scientific discovery, not only for the peacefulness, but because I know that what is written during this time will be used in several ways. For the students, the entries offer a record of the processes they followed and a celebration of their successes; for me, the journals are a valuable tool for assessment. I am able to review the questions that were asked and gain further insight into the students' thinking.

It is at this point that teachers can make an early assessment of question quality. Are the students asking the kinds of questions that will lead to meaningful investigations? Often, at this still early stage, the questions are somewhat shallow. Some are merely "yes" and "no" type questions. "Will sugar dissolve in vinegar?" Others are "Can I . . . ?" type questions. "Can I make an aluminum foil airplane that will fly?" Although these may be interesting questions, I know that through modeling, direct instruction, and the use of student examples, I will eventually guide the children into asking questions that are more complex.

Questions that compare, such as "Which dissolves sugar faster, cold water or warm water?" lead the children to examine variables in their research which requires more careful planning and can lead to other questions. "What if . . . ?" questions lead to varieties of possibilities as the children design their own experiments. What do I expect will happen if I do this or if I do that? The children use what they know to find what they do not know. But it is difficult to accurately assess the quality of a question by merely examining the question itself. A great deal can be learned about question quality by looking at the *results* of a particular question. Seeing where a question takes the investigator is the true test of the quality of that question.

Later in the year, these discovery box inquiry periods take a different form as the children progress toward more in-depth investigations resulting from the growing sophistication of their questions. Increasingly, children make their own choices for each inquiry period. Some branch away from the boxes as they explore their own more extensive pursuits. These investigations originate from the boxes, KIC journals, or the experiments and investigations begun by others. In Chapter 8 we will see other sources as inspiration for student investigations.

Management of subsequent inquiry periods is made easier with the Plan For Inquiry Period form (Figure 4-4). A day or two before the inquiry period (after signing up for a box, if one is to be used), each student completes this form. The student specifies what he or she intends to do; either a discovery box, contract (see Chapters 5 and 8), or other investigation topic. If a discovery box is to be used, the box is checked to see that needed materials are present. If a contract or other investigation topic is to be worked on, the student makes certain that needed materials are identified and available and, if not, who will be responsible for obtaining them. The goal is to avoid idle time during an inquiry period because of insufficient materials. Placing the responsibility on the student helps to transfer the ownership of the experiences away from the teacher.

The Book of Discoveries

Following the first discovery box period the class is introduced to the Book of Discoveries (see Figure 4-5). This binder, which resides in our class library, gives each student the opportunity to publicly record their own discoveries for others to read. The entries stay in the book even after the year has ended, which means that students can read about discoveries from previous years. Plus, as students make entries into the book, they know that students next year and in the years to come will be

Plan for Inquiry Period

Name ______________________________ Date ____________

Next Inquiry Period __________________________

Day and Date ______________________________

BOX, TOPIC, or CONTRACT to be used: ____________________________________

(If CONTRACT, list contract name and number) ______________________________

Testable Question you will be trying to answer: ___________________________

____

____

Materials to be used: ____

____

____

____

Are the needed materials available in the classroom now? YES NO

If "NO," which materials are needed?____________________________________

____

____

____

List any special needs necessary for your investigation. ______________________

____

SUBMIT THIS FORM TO YOUR TEACHER AT
LEAST 24 HOURS PRIOR TO INQUIRY PERIOD.

FIGURE 4-4 *Plan for Inquiry Period*

Scientific Discovery

This discovery was made by ______________________________
student scientist(s)

on __________________________
date

Discovery

____

____

____

____

____

The following steps were taken which led to this discovery.
(Please include specific details such as measurements, quantities, and one or more sketches or diagrams.)

1. ____

____

2. ____

____

3. ____

____

4. ____

____

sketch

Confidence of Accuracy (circle one)

HIGH LEVEL OF CONFIDENCE 10 9 8 7 6 5 4 3 2 1 0 LOW LEVEL OF CONFIDENCE

FIGURE 4-5 *Book of Discoveries Entry Page*

Replications

I. *The experiment/observation described on the front side was repeated by*

____
student scientist(s)

on ____________________.
date

Results: ____

____

☐ Results Similar To Original Discovery	☐ Results Inconclusive	☐ Results Different from Original Discovery

---

II. *The experiment/observation described on the front side was repeated by*

____
student scientist(s)

on ____________________.
date

Results: ____

____

☐ Results Similar To Original Discovery	☐ Results Inconclusive	☐ Results Different from Original Discovery

---

III. *The experiment/observation described on the front side was repeated by*

____
student scientist(s)

on ____________________.
date

Results: ____

____

☐ Results Similar To Original Discovery	☐ Results Inconclusive	☐ Results Different from Original Discovery

FIGURE 4-5 *(Continued)*

able to read what is written. The back of the form provides space for other students to record the results of later attempts to replicate the experiment.

The Book of Discoveries adds to our sense of community. As scientists, the children are expected to report to one another. The discovery book is one forum for doing that. Our community consists not only of this year's students, but of all the students who have been part of our class. Students from previous years have much to contribute to future students. Too often we overlook all that is shared by our students.

I am always amazed when students return to my classroom several years after leaving it. One of their first questions is often, "Is my entry still in the book?" Even as middle and high school students, they have a sense of pride in what they contributed earlier. In a way, they are still a part of our class and always will be.

The Book of Discoveries is also another source of authentic reading. Some students may be inspired by what others have written. In attempting to replicate experiments, as students follow the procedures written by others, they find that clear writing is essential to scientific inquiry. This realization sometimes motivates students to improve their own writing. The student authors of entries in the book will not be here next year to explain what they are trying to say. Their writing must say it all!

Five

Outdoor Education

Nature writer Edwin Way Teale once told the story of watching a young boy who had found a dead monarch butterfly in a pile of litter beside a street. The boy stood entranced, bending forward, oblivious to all around him. Mr. Teale remembered, "It seemed as though I were looking at myself when young. A door was opening for him, a door beyond which lay all the beauty and mystery of nature."

Watching and listening to our students outdoors can be just as pleasing.

"Look at this big hole!" exclaims Sean. "Maybe a beaver made it."

"No, it couldn't be a beaver!" responds Rebecca. "There isn't any water. I think it was something else."

"Look!" cries Brandon. "There's a tree that fell down. Maybe it used to be in the hole."

Figuring out what goes on in the world around them comes naturally to children. They enjoy gathering data through their own observations and explorations. Merely watching children of any age will support this premise. Kids love doing what we call science (although they may call it any number of other things). It also seems that virtually all children like going outdoors. When asked to list their favorite things to do, most include some kind of outdoor activity high on their list. Certainly in a school setting, going outside is a highlight of any day.

Real science in the hands of children can be a powerful way to motivate them to do well in other areas. Imagine combining the excitement of inquiry science with the call of the outdoors. Outdoor education can include both! As an enhancement to inquiry science back in the classroom, outdoor explorations can easily become a vital part of any program of studies.

The roots of outdoor education (or environmental education, as some call it; the terms sometimes blur) can be traced back to the 1800s. Naturalist Louis Agassiz's motto was "Study nature—not books." This philosophy seemed to dominate the period. Teachers were encouraged to engage their students in firsthand experiences with nature as opposed to merely learning from textbooks. Thoreau, John Muir, and Theodore Roosevelt emerged from this kind of nature study. Since that period, the concept of outdoor education has evolved and changed.

For some, outdoor or environmental studies primarily involve topics like rain forest depletion, recycling, global warming, extinction of species, pollution, and energy conservation. It is difficult for children to assume ownership of such weighty problems. A partial awareness may be important, but student action (beyond letter writing, collecting money, or poster making) is limited.

Others see an environmental awareness as more grassroots oriented. The global issues will be tackled in due time. Local ones are seen as prerequisites. Issues like diversity of life in a lawn, observations of migrating butterflies and birds, cataloging of life in a school yard, monitoring of erosion and runoff, local stream quality and bioindicators, are issues even primary children can internalize. With a growing familiarity come questions; with those questions come the need to observe and investigate. The world beyond the school's door is a well-stocked laboratory that is inviting, always present, and ever changing.

The concept of educating outdoors is catching on at a rapid rate. Schools in a wide variety of environments are incorporating school-yard settings as a natural and authentic reason to go outside. Organizations of outdoor educators are growing at phenomenal rates in numerous states. For example, the North American Association for Environmental Education (NAAEE) now has forty state affiliates and is part of a fifty-five-nation network. Organizations like the Maryland Association of Environmental and Outdoor Educators (MAEOE) are reporting an annual rise in memberships. MAEOE, which began in 1985 with thirty members, swelled to a membership of over three hundred in 1998. A growing percentage of groups like MAEOE is comprised of classroom teachers who are extending their lessons beyond the classroom walls. Increasingly, outdoor education is being seen as something other than a fringe extra to be scheduled only on Friday afternoons if time permits.

Teachers cite a variety of reasons for taking their children outside. Certainly environmental awareness is one of them. But as an enhancement to inquiry science, forays outside allow for exploring, questioning, investigating, and discovering. Given the time and place, children will examine bushes, lawns, weeds in sidewalk cracks, trees, meadows, streams and ponds, and even forests. The questions that evolve from these kinds of activities are endless.

School-yard habitats can run the range from an asphalt play area to acres of forests and meadows. As different as these outdoor areas may be, they all share several important characteristics. First, they all contain some form of life that can be observed by children. Weeds, lawns, flowers, bushes, and trees are all miniature ecosystems full of life just waiting to be discovered. Second, every school yard provides opportunities for children to experience the weather. From watching clouds to tracking shadows to feeling and measuring the wind, each school yard is a portal to the environment and each school yard is linked to all of the others because of it. Third and possibly most important, each school-yard habitat belongs to the children. The students assume ownership of the space. This gives them not only a sense of pride for what is there, but also a sense of responsibility to care for it. How an outdoor facility is developed and used depends, of course, upon available resources. But *any* outdoor setting can be used to enhance inquiry.

Getting Started

The first step in establishing an outdoor classroom facility is to assess what is already there. This is the easy part. Virtually any outdoor setting will do. What any school has at the outset can be used by the students immediately. The only requirements are that the available facility permit safe, student exploration, regardless of how modest it may be. The discovery and documentation will naturally follow.

Most schools have one or more trees scattered about the property. Trees have traditionally been great friends to children. They provide shade, are great to climb (although climbing a tree at school should probably be discouraged), are big yet nonthreatening, and can always be relied upon to be there tomorrow and the day after. Trees in the school yard should be numbered. This provides several benefits. Numbered trees can serve as reference points. Describing a discovery that occurred twenty paces north of tree #04 is a clear way of precisely locating that position. Numbered trees are also useful when students are asked to render the space in writing. Students get an understanding of ratio and scale when they transfer an actual measurement between trees to a scaled-down measurement on a drawing. Once some or all of the trees in a space are numbered, an initial activity by the children can be to draw a map (either to scale for older children or simply freehand for younger ones). Numbered trees can also be referred to for identification. It is recommended that trees *not* be identified with signs. What fun is it to be told everything outright when there are books inside to help children find the answers?

There are many ways to number trees. One temporary approach is to write numbers on index cards and attach them to the trees with tacks or yarn. A more permanent way is to paint numbers directly on the trees with exterior paint. These numbers will last several years and are resistant to vandalism. I have used a router to inscribe numbers on boards (of about four inches square), which I then attached with small nails to the trees. Using a ladder to attach the numbers out of reach will encourage the children to look higher in the trees and will also minimize vandalism.

In a school yard with only a few trees, numbering each would be important. However, in a wooded area, numbering trees at scattered locations works well. The locations of any other trees can be identified by referencing the numbered ones. It is a good idea also to use a wide range of random numbers rather than sequentially counting from 1. Prime numbers, bus numbers, factors of ninety-six—whatever seems interesting. The numbers will take on their own significance as they become familiar parts of the school yard.

Adopted Trees

Getting the children acquainted with your outdoor classroom is a great way to start the year. If you have several trees numbered, a walk around the grounds with the children looking for numbers will encourage them to look up into the trees, leading to numerous possible observations. Following the walk, the children can share with the group what numbered trees they observed. I like to then give the children a map (not to scale) with subject trees marked but not identified by number. The children go back to find whichever trees they may have missed and fill in the numbers on the map. When finished, the maps give them a complete overview of the numbered trees.

This is the time for the children to adopt a tree. They are permitted to select any tree (either numbered or not) and are asked to complete a tree-adoption form. In their dialogue journals, the children will jot down observations, changes, and interesting facts throughout the year as we return to visit. Initially, they are to provide the precise location of their tree on the back of the adoption form. A reader should be able to locate the tree by reading the description. For example, one student's tree may be located fourteen paces northeast of tree #17. An understanding of nonstandard measurement as well as cardinal directions (north, south, east, west) is required.

Sweep Nets

Regular visits to the outdoor classroom have provided students with one adventure after another. Our excursions elicit lots of questions to add to the question board, and we always return with many items to display on our nature table.

In the late summer and early fall, sweep nets are a great way to gather insects and spiders from the fields. Even mowed grass contains a variety of otherwise unseen critters that can be gathered in a net. Sweep nets are cloth nets on three-foot-long poles, available from science supply companies or easily made out of coat hangers and nylon stockings by parent volunteers (see references at end of this book for sources of plans for making nets). Sweeping the net back and forth while the student is walking forward will yield an amazing array of life. Ideally, it is best to sweep across unmown grass (requesting a part of the school lawn go uncut for student research purposes would provide impressive results). We call our uncut area our "meadow." When sweeping in the meadow it is important to document what is found, either in the student journal or using some other tool (see Figure 5-1 for a sample meadow-sweeping data sheet).

The fall is just the time to think about spiders. Sweep nets used in September and early October yield many spiders. The children seem to have a natural fascination for these creatures. We have set up predator habitats (small clear containers with twigs and grass clippings) in which spiders have woven webs. The children have witnessed prey being captured and eaten. We have also seen spider eggs hatch with what appeared to be hundreds of baby spiders climbing around the enclosure. With some thirty-five thousand species of spiders worldwide, quite a few will be outside any school. Once some of the children are hooked on spiders, they will be observing and recording in the classroom and finding books about spiders in the library. This may not be the time for a classwide unit on spiders, but it might be just the time for a reluctant reader to want to find out more from a book. It is also the right time to read aloud to the class *The Lady and the Spider* by Faith McNulty.

The Salamander Census

Other animals easily found in the fall are salamanders. Any wooded location with fallen logs and branches on moist ground will support salamanders. These little critters are easily handled, are safe, and are fascinating to capture and watch. It should be communicated to the children, however, that capturing salamanders is not just a fun activity. Salamanders and other amphibians are disappearing worldwide at alarming rates. Important data can be gathered to help track the salamander population at your location. Reading *The Salamander Room* by Anne Mazer helps children to think about their responsibility to wildlife.

A three meter by three meter square can be roped off in an area where salamanders are known to live. By carefully examining under rocks and logs, salamanders can be collected. A data sheet is used to record salamander observations. Information includes location, temperature, weather, time of day, square meters of search area, number of salamanders found, and average length. Not only do these observations lead to many student questions, they also involve authentic uses of the math skills encountered back in the classroom. Later in the year the census can be repeated to look for changes that might be seasonal. (It is *extremely* important that the children release the salamanders soon after the census data have been collected.)

In future years, the ongoing salamander census might reveal significant changes in population. This would be important for students to be aware of and to note. Drawing upon the salamander research of earlier students and trying to make sense of new data adds to the sense of community being created in the classroom. It underscores common effort as students work together on projects that are important and authentic.

In the Meadow

Date ____________________

Your Name ____________________________

Your Partner's Name ______________________

Today's Weather _______________________

Use your net to sweep the meadow.
Look to see what you have collected.

Make a drawing of one of the creatures in your net.

Describe your creature.

COLOR ________________ NUMBER OF LEGS __________

DOES YOUR CREATURE HAVE WINGS? YES ___ NO ___

DOES YOUR CREATURE FLY? YES ___ NO ___

DOES YOUR CREATURE JUMP? YES ___ NO ___

MEASUREMENTS: length _________ width _________

ABOUT HOW MANY OF THESE CREATURES ARE IN YOUR NET? _____

WHAT IS THE NAME OF YOUR CREATURE?
(Make one up if you do not know!) ________________

Write three or more questions you can think of about your creature.

1.

2.

3.

Before leaving the meadow, remember to release the creatures in your sweep net.

FIGURE 5-1 *In the Meadow Data Sheet*

Biodiversity Day

Most of the children have probably heard about the depletion of the rain forests and the loss of species as a result. It has been estimated that our planet is losing twenty-seven thousand species annually. That's seventy-four per day, or three species per hour. This represents what may be the biggest extinction crisis since the disappearance of the dinosaurs. For some children, this may be scary. When will it happen to us, they may ask? (How perceptive so many of these children are!) But this is a crisis over which the children have no control, nor can they do anything significant to alter.

It has also become apparent to these children through their outdoor explorations that their school yard is teeming with life; probably much more than they ever imagined. Some may wonder if the life around their school may likewise be affected by what is happening elsewhere. This is something the children *can* do something about and make a significant impact upon. Although we work on this project all year, it is introduced and highlighted during our first Biodiversity Day.

I tell the students about our five-year mission to document life in the outdoor classroom. Since so many life forms exist there, the efforts of everyone are needed. Insect books, field guides, and books about birds and flowers and trees are borrowed from the library and taken to the outdoor classroom. We place all of these reference materials on a big blanket along with empty containers, hand lenses, and microscopes. The sweep nets are there and any other tools and materials that might help the children with their mission.

The challenge is to locate a plant or animal in the outdoor classroom and identify it using a reference book. This requires the children to use whatever clues may be available. For example, if some students think that they have located an insect that is shown on a map in a field guide as living in the Pacific Northwest, but the students are in the Southeast, the chances are that they have misidentified the insect.

To help with this project, we practice using field guides before going into the field. Students are challenged to identify plants and animals for which a definite identification has already been made. This becomes almost a game as the children discuss the clues and the evidence. It may seem to be an activity reserved for older children, but even primary-aged youngsters show a knack for discriminating between species, especially when given assistance. We have been successful with science buddies, fifth- and second-grade students paired to work together. This activity has also worked well for pairs of eighth- and fifth-grade students. The excitement and authentic challenge combine to bridge the age gap.

Listening to their excited conversations allows me to see how the two age groups can learn from one another. When an eighth-grade class came to help us on an early fall Biodiversity Day, they brought new insight and ideas. Since it was fall, crickets filled the sweep nets. While my fifth graders were the experts at sweep net technique, the older students were more skilled observers. After placing a cricket in a small plastic container, two girls began to study the insect.

"Its leg is crooked," commented the fifth grader.

"It isn't crooked; it's bent to help it hop. Look. The leg is bent at a forty-five-degree angle," the older girl pointed out. The two girls spent a long time completely engaged in insect observation.

Discoveries are documented on the Life at the Outdoor Classroom data sheet (Figure 5-2). The students describe what they have found, make a sketch, and include a reference. The goal is to convince the Outdoor Classroom Committee, a committee made up of any combination of students and teachers or simply the classroom teacher, that the plant or animal was actually found and identified correctly. If the Outdoor Classroom Committee is convinced of the authenticity of the documentation, then the form is stamped, given a number, and placed in the official Outdoor Classroom Life book, to remain indefinitely. Whenever that plant or animal is seen in the future, the original student who first discovered it is cited as such. A student who discovers and documents a dandelion will forever be given credit for making that discovery. Their names will be in the big book for years to come.

This project has served to document a lot of the species in our outdoor classroom. These data will become increasingly invaluable as the years progress as a record of life. Regardless of how many species have been identified in the past, there will be plenty of different ones for students to find in the future. This is one project that will never be complete.

More Outdoor Experiences

Throughout the year, visits to the outdoors will lead to more and more observations and questions. We always take along field guides and the students' dialogue journals, and carry the outdoor classroom kit of supplies. Included in the kit are latex gloves, containers of varying sizes for specimens, plastic sandwich bags, rulers, tweezers, and hand lenses. (A first aid kit is also a must.)

The winter months can be especially interesting. With underbrush at a minimum, parts of a wooded area are exposed for exploration. Poison ivy, though not totally absent, is dormant enough to pose little difficulty.

One of the most exciting parts of visiting our nearby wooded area is the discovery of bones. Imagine how thrilling it is to stumble upon a location in the forest where the skeletal remains of an animal are waiting to be found.

The cry goes up. "Mr. Pearce! Look what I found."

"Oh gross!" say some.

"Wow! Cool!" say others.

The reactions vary, but all the kids rush over to see what is there. My role is to

Life at the Outdoor Classroom: Plants and Animals Observed and Identified

Student Name ______________________ Date of Observation ______________

Name of Life-Form ______________________ Circle One: PLANT ANIMAL

Location of Observation ____

DESCRIPTION

SKETCH

VERIFICATION: Book ______________________ Page ____________

Other students who also observe this life-form are invited to record information on the reverse side. Please include student name, date, and location of observation.

FIGURE 5-2 *Entry Sheet for Outdoor Classroom Life Book*

enforce some safety rules without inhibiting their enthusiasm. Usually I model a questioning process.

"What do you think it is?" I ask.

"Bones," the discoverer replies.

Since the whole class is gathered together, we share the experience. Step by step, I take them through a series of questions, asking where the bones might have come from, what might have happened, and what we should do with them. This takes time, but it is important. Science is a story to be told, and nowhere is this made more clear than in animal stories. Most of the bones we have found were those of small mammals such as rabbits and squirrels. Because of the frequency of these discoveries, I prepare the children by telling them about what *not* to do when bones are encountered. They know not to touch the bones or pick them up with bare hands. Any number of dangerous bacteria are on and in the bones. There are safe ways to handle and preserve the find.

First, if bones are found and we decide we want to take them back to the classroom, we put them into a plastic bag using the latex gloves from our outdoor classroom kit. Then, when we get back to the room, there are several ways to make the bones safe for display. One way is to boil the bones for an hour or so. The problem with boiling is the smell! The preferred way is to soak the bones in a solution of bleach and water. Though effective, these methods are not to be performed by students.

Other bones have been found as part of owl pellets. These are the regurgitated remains of very small animals eaten by owls. In the pellet are the bones and fur of the animal that could not be digested by the owl. These animal remains are usually those of mice or meadow voles, but can also be those of birds. When supplies are plentiful, owl pellets are fun to pick apart in the field with small sticks that can be discarded in the brush. For a more formal investigation, the pellets can be returned to the classroom in plastic bags. They can be made safe for student use by placing them in a microwave oven for twenty to thirty seconds or in a 250-degree conventional oven for fifteen to twenty minutes. Nevertheless, students should always be reminded to wash their hands after handling even properly prepared bones. Elementary students can take apart owl pellets using wooden pointed skewer sticks that can later be discarded. Owl pellets can also be purchased. Bone charts are available that assist in identifying the animals in the pellets. (See Chapter 8 for some excellent Web sites with information about ordering and examining owl pellets.) Unraveling the mysteries that hide within each owl pellet lead to many more questions to consider.

Similar investigations can begin on a smaller scale right on an asphalt playground. Insect remains fascinate children. It sometimes surprises students to discover that predation occurs on a smaller scale right beneath their feet. Inquiry is about student questions and observations, and those questions often start in the places the children visit most often.

Winter is also ideal for outdoor explorations because of snow. Though a forested area is not required for the study of animal tracks, if a wooded area or field *is* available, the stories told by tracks in the snow are fascinating.

Teacher/Student Contract

Contract Number ______ Date ____________________

________________________ agrees to work on the following long-range assignment:

WRITE A FIELD GUIDE

to be completed on or before ____________________________

The classroom teacher agrees to provide a reasonable amount of class time by exempting the student from selected classroom assignments.

Completion of this contract (will/might/will not) require additional time at home.

The following provisions apply in the completion of this contract:

- The student will identify five or more plants or animals in a portion of the outdoor classroom.
- The student will write a field guide for that portion of the outdoor classroom in which the plants or animals are described. Possible titles for the field guide might be:

  Plants of the Forest of Integers

  Animals of the Pond

  Insects of the Meadow

  or another title chosen by the student.
- The Field Guide worksheet will be used to display the information about the plants or animals.
- A cover for the field guide will be prepared which includes the title and author.
- A title page will also be prepared which includes the title, author, and date of publication.
- A bibliography will be included in the field guide.
- The final version of the field guide will be free of spelling errors.

The student agrees to do his/her best work on the completion of this contract.

____

STUDENT

____

PARENT

____

TEACHER

____

CLASS

FIGURE 5-3 *Teacher/Student Contract for Write a Field Guide*

When the weather is cold, an animal's tracks will remain for days. Children love to find the evidence of an animal's presence, especially since most animals run for cover long before a noisy class comes into an area. In the snow, even the noisiest of classes will find tracks left behind by busy animals.

The early winter is the perfect time to teach the children about tracking. Many books show the prints of animals that might be found in your area. Deborah Parrella's *Project Seasons*, published by Shelburne Farms in Vermont, has an especially useful section about tracking. The children will learn about the patterns, prints, and the places tracks are found and will learn how to identify tracks. One winter, while exploring our nearby forest, my class and I came upon deer tracks that were quite clear. We decided to follow them and came to a secluded place in the woods where there were numerous indentations in the snow, apparently the place where the deer had slept the night before. By examining the tracks and observing where they went we were able to tell a story about the deer. We had been to that area of the forest many times, but had never seen such evidence of the presence of the deer. The snow revealed much about the life in our forest.

Throughout the year, we gather data to write our own field guides. In the fall, the children became familiar with *Peterson's First Guides*. Later, the children are invited to create guides for their own outdoor area. Guides to birds can be written at any time, especially when numerous birds are visiting nearby feeders. Guides to spiders or insects can also be written. We use a contract (Figure 5-3) that clearly shows what is expected of the students creating the guide. A sample field guide worksheet is shown in Figure 5-4.

Throughout the year students can record weather data in their outdoor location. From simple temperature readings to more complex wind, barometric, and humidity readings, the children use instruments to gather information about the atmosphere. Our school weather station is especially popular when snow is coming. We have also been tracking the water temperature of a nearby pond. Equipped with thermometers, students take readings from around the pond, then find the average back at the classroom. The monthly readings are charted to see how high and low the temperatures fluctuate.

Visits to our outdoor classroom continue throughout the year. As the children share experiences outdoors, class community and cohesiveness is strengthened. In addition, visits outside help our children to know and to come to love the plants and animals they find there. Touching the affective domain in the children will enable them to not only understand, but to care.

Inquiry, the outdoors, and children are natural companions.

> There is indeed a chasm which separates man and animal, and . . . if that chasm is to be bridged, it must be man who does it by means of understanding. But before we can understand, we must know; and to know, we must love. We must love life in all its forms, even in which we find least attractive.
>
> *Jacques-Yves Cousteau*

Field Guide

_Of_ ____________________

Location

(circle) ANIMAL PLANT

Author ____________________

COMMON NAME ____________________ SCIENTIFIC NAME ____________________

DESCRIPTION ____________________

____

____

HABITAT ____________________

WHAT IT EATS and/or WHAT EATS IT ____________________

____

____

BEHAVIOR ____________________

____

____

SIZE (measurements) ____________________

____

INTERESTING FACTS ____________________

SKETCH

____

____

PAGE________

FIGURE 5-4 *Field Guide Page*

Inquiry in Action: The Search for a Natural Antibiotic

Away from school, children do not generally separate their thinking into subject areas like language, math, and science. They integrate and use the academic skills they know to work on whatever task is at hand. Educators, however, tend to segment the school day. With the autonomy of inquiry, it is often the children who make connections between two or more seemingly different areas of study. The combination sometimes proves to be innovative and entirely unexpected.

Kayce was struggling with her research on antibiotics. This topic may at first seem advanced for elementary students, yet it proved intriguing for several boys and girls after meeting with scientists Barb Bradley and John Hejna of Becton Dickinson, a biomedical manufacturing corporation. The students learned that antibiotics are chemicals used to kill bacteria. With bacteria cultures, the students were able to experiment with a variety of antibiotics. The process was simple. First, an agar plate is streaked with a bacteria culture. Then, a small paper disk (about 5 mm diameter) is immersed in any liquid to be tested for its germ-fighting capabilities. The disk is then placed on the streaked plate. After twelve to eighteen hours, the plate is examined. The plate will be covered with the bacteria. If the antibiotic is effective, a clear zone will be around the disk where the bacteria's growth has been inhibited. Once the children understand the process (and the necessary safety precautions of such research) they are able to go on and conduct their own research with antibiotics.

Kayce was curious about the sources of antibiotics. She knew that drug companies manufacture medicines that fight bacteria. But she wanted to know if she could find other sources. During the time Kayce pondered the possibilities, our class made several visits to the woods in our outdoor classroom.

Kathleen became interested in Kayce's inquiry. It was she who, a year earlier, had noticed bacteria being attacked by mold in her bacteria culture. She had shared her experiences with Kayce, who wanted to continue with the research. The girls knew that rain forests were sources of a variety of medicines. It was here that the connection was made. The research with bacteria in the classroom might be combined with visits to the outdoor classroom, where natural antibiotics might be found. Once the hunt began for a natural antibiotic in our own forest, Kayce and Kathleen and others who joined them were led on a journey full of frustration and surprise.

First they tried spring onions. The culture plates were prepared and the onions ground to provide juice for the paper disks. The next morning the girls were excited to find a clear zone around the disk. The onion juice was indeed an antibiotic and Kayce and Kathleen were proud of their discovery.

"Yes, but how do you know the paper disk alone didn't stop the bacteria from growing?" I asked. "Did you include a control?"

Kayce and Kathleen looked at each other and realized that their experiment was flawed. They had learned about the importance of controls but only now saw how necessary a control can be in authentic research. This was where the frustration began.

The girls went to the back table and found some remaining onions from the day before and used them to replicate their experiment. But in addition to the disk soaked in onion juice, they included a disk that contained no liquid as a control.

The next morning we looked at the bacteria. It had grown right up to the edge of both disks.

"What happened?" Kayce asked. "We didn't expect the control to work, but why didn't the onion juice stop the bacteria?" Kathleen was stumped as well. By now the class had gathered around to see what was going on.

"Maybe the onion juice was old," someone suggested.

"Or maybe you used a different part of the onion plant," someone else said.

Kayce and Kathleen were determined to try again and were eager to go back to the woods for more onions. Unfortunately, we had to wait out several rainy days before we could return.

A week and a half later the girls gathered a fresh supply of spring onions. Kayce carefully retraced her steps and set up the experiment as before. But once again the onion juice failed to stop the growth of the bacteria.

"Something is different from the first time we tried it," Kayce said, "but I can't think of what it is!"

Many suggestions came from other interested students.

"Maybe you got onions from a different place in the woods," one person suggested.

"What time of day did you pick the first onions?"

"What about the weather?" someone else asked. "Maybe since it rained, the onions are weaker."

Kayce was frustrated because she could not remember all of the details of her first success. "I wish I wrote it all down," I heard her mutter. "I could just go back and see what I did differently. From now on I'm writing *everything* down!" Often the best lessons are the ones the kids teach themselves.

Kayce and Kathleen tried dozens of plants from the forest with no success. Then one day we were in the woods searching for an early blooming plant called bloodroot. Its distinctive white flowers last only a week or two and are a harbinger of spring. Legend has it that Native Americans used the orange-colored juice of the root for a variety of skin ailments. Kayce and Kathleen thought that this might be their natural antibiotic.

Carefully they processed the root to extract enough juice to wet a paper disk. After following their procedures, the girls were anxious to see the results the next day. What excitement there was when they found a well-defined clear zone around the disk soaked in bloodroot juice. "And the control disk," Kayce proudly announced, "has bacteria growing all around it right up to the edges."

Kayce and Kathleen told their story at the Kids' Inquiry Conference that May. Their perseverance paid off, and their discovery was an inspiration to everyone.

SECTION II

Tending the Fire: Sustaining Inquiry Throughout the Year

As kids are exposed to inquiry science, a level of facility is eventually reached in which the children begin seeing more and more connections and patterns. Each new experience elicits new questions and investigative possibilities. The campfire of inquiry begins to take on a life of its own, sustaining itself as it burns ever brighter.

Thus far, the children have been guided through a process of questioning and discovering. They have been provided with teacher and classmate modeling to develop their own questions and the question board on which to record them. Those children who may have launched their own investigations early in the year have shared their findings through informal presentations and entries in the Book of Discoveries.

Students have been reading about the adventures of past years. Journals of the Kids' Inquiry Conference have been used as reading materials, telling stories about the inquiry experiences of previous students. (The Kids' Inquiry Conference and its journals are discussed in Chapter 9.) The folders in the discovery boxes tell of experiments and observations performed by former students. Increasingly, the children see themselves as part of a community in which they will eventually contribute their own stories and discoveries for later students to read.

The role of the teacher thus far has been that of leader and guide. It was the teacher who gathered the wood and prepared the fire. The teacher provided the environment in which questions have flourished and possibilities have been imagined. The teacher has been a scientist, modeling through think-alouds and posing provocative questions before the class.

As the year progresses, however, the roles that the children and teacher play begin to shift. Increasingly, the teacher guides from behind, still providing opportunities and materials, still monitoring and assessing, but not always determining where the children may go. Responding to past experiences and their own curiosities and questions, the children begin to take on a greater degree of independence. The time must come for the teacher to relinquish control and allow the children to think of themselves as scientists, to actually *be* scientists.

The wood on the campfire cannot be forced to burn. All that can be done is to provide the proper conditions. So it is with the children. We cannot force them to be curious or to ask questions or to pursue investigations or to document their discoveries. Yet, by creating the suitable conditions, all of this will occur.

It is sometimes hard to know when to stop stirring the fire and when to let it burn on its own. The time likewise comes for the teacher to trust the scientific instincts within the children and for the children to have confidence in themselves as scientists. Through it all, each new experience offered to the children becomes a new opportunity for inquiry.

Six

Using the Established Curriculum

"This is fifth grade. Kids shouldn't be using batteries. We covered electricity in fourth grade."

"You can't let kids decide what to do. They'll just goof off and do nothing."

"Our school year is too busy to overlap topics from one year to another. Plus, the children will get bored!"

Have you ever heard similar comments in the faculty workroom? There is much reluctance on the part of some educators to see science as something children do, rather than simply as something they know. One of the major goals of inquiry-science instruction is to motivate the children to actively question and investigate. For many teachers, this is also a major problem. These teachers assume that inquiry science is too difficult to achieve in the classroom. The demands for materials and the loss of teacher control seem more than can be managed. Furthermore, many educators assert that children are not able to explore independently and are not capable of asking quality testable questions until secondary school.

Inquiry science involves far more than merely giving children materials and books. Inquiry is a process that demands preparation. The children, although natural scientists, must understand the process of science. It is imperative that students view science as a systematic and logical way of expanding knowledge and understanding. The children must be able to make connections, notice patterns, and go on to ask the quality questions that lead to meaningful investigations and discoveries. Yet, inquiry science is not part of the standard science curriculum that many teachers must teach. How can teachers reconcile their commitment to inquiry with curricular constraints? The answer lies in using a curriculum as a beginning—not an end.

The science curriculum can be a valuable tool for inquiry. Not only does an effective curriculum provide the content so vital for prior knowledge, but it also refines the process skills students must have in order to engage in their own investigations.

Many teachers view their science curriculum as a lifeline, guiding their students (and themselves) into areas they might not otherwise venture. Both teachers and students can learn much from the curriculum. There are also times, however, when the curriculum is too confining, demanding that both students and teachers do things they neither care about nor find relevant. Some teachers have even claimed that their science curriculum takes time *away* from real science. When a rigid step-by-step curriculum written by people who do not know the children dictates what occurs in a science classroom, students must fit the curriculum rather than the curriculum fitting the interests and needs of the student.

As with so many of the tools available to the classroom teacher, the curriculum is there to be used in whatever way possible, not only to provide the content and process skills of science, but also to enhance what the children are able to do *beyond* the curriculum itself.

Any teacher can utilize the curriculum to encourage student inquiry in the classroom. Looking over the shoulders of students engaged in inquiry-based investigations might help show how curriculum requirements enhance the inquiry process.

A Cavern Investigation

Joe, Brad, and Mike were in my fifth-grade class, which was studying earth science. In small groups, they had participated in several hands-on activities about weathering and erosion topics from the school district's science curriculum. One activity involved the pouring of acidic water over limestone chips to see how weathering might occur. The experiment demonstrated how the water breaks down the limestone in the process of chemical weathering.

Earlier in the year the class had visited a quarry in which limestone was mined. The students were told about limestone, the materials it was made of and how it was formed. They were told about the uses of limestone and about how this rock was being attacked by acid rain.

The boys had also been working with a discovery box in which there were a variety of materials from the earth science unit: jars, vinegar, pH paper, limestone chips, and other rocks and minerals. At one point, the boys placed some limestone in a baby food jar, covered the chips with vinegar, and then placed a balloon over the jar. They watched as the balloon began to inflate with carbon dioxide gas, released as the acid interacted with the calcium carbonate in the limestone. Perhaps as a result of class discussions or informal conversations with me, the boys made a connection with all that they had heard and seen. *If*, they pondered, there was limestone under the ground to be mined, and *if* limestone was affected by acid, and *if* rain was acidic, wouldn't the limestone under the ground eventually be dissolved by the acid in the rainwater?

The earth science unit was long over as they asked these questions, but it had never come to an end in the boys' minds. They became obsessed with these questions, questions that would never have occurred to them had the science curriculum not been there as inspiration.

Joe, Brad, and Mike then went a step further. Based on what they knew, they wanted to find out more. What *does* happen when acid reaches limestone under the ground? Joe found a book and showed the others about the formation of caverns. Brad said he had visited a cavern once and he thought it had been formed because the rock had been dissolved. The three decided to make a model to see for themselves. Their question was: "How do caverns form in limestone?"

As I witnessed the dynamic of the problem shift from a teacher-initiated activity to a student-owned investigation, I knew that my role was to provide space for the ideas to grow and evolve among the boys. Through the curriculum, I had set the process of inquiry in motion. The boys had now taken over control of the experience.

We found some powdered limestone that the boys thought might dissolve faster than the limestone chips. Their concern was that the process of dissolving limestone might take a long time, so they wanted to speed up the process in their model. In a plastic jar with drainage holes in the bottom, the boys placed a layer of gravel covered with the powdered limestone. On top of the limestone they fit a screen and placed soil and gravel above it. Their model was ready.

For several days the boys carefully poured a solution of vinegar and water into their model. They used a similar proportion as was used in the class experiments (although much discussion was held between them as to the exact amounts to use; too much vinegar would constitute an unfair test, while too little would take too much time to produce results). They expected to see the powdered limestone begin to dissolve. After a few days, there were pockets in the limestone that were left behind by the percolating acid. The boys measured the pH of the solution going in and compared it with the pH of the drained liquid in the bottom of their model. The pH had risen, they proposed, because of the limestone.

What is my role in this process? I suggest initial ideas, pose challenging questions and dilemmas, facilitate class discussion, and provide the needed materials and the time to use them. The curriculum assists me and my students by providing a framework within which inquiry can grow.

Links with Standards

Topics of inquiry in the classroom can be traced back to a variety of origins. Established curricula can be used to help inspire continued curiosity among our students. Furthermore, the curriculum can be used to *support* our use of time and resources for inquiry.

Most curricula today are written using a variety of standards as guides. Organizations like the American Association for the Advancement of Science, which has published *Science for All Americans* as part of Project 2061* (see note at bottom of page 66), has aided the writing of curricula. An increase in scientific literacy is directly linked to the inclusion of inquiry in the classroom.

The National Science Education Standards, published by the National Research Council, has become a blueprint for local school systems to improve science curricula and assessment materials. The standards are divided into six strands: science teaching standards; standards for professional development for teachers of science; assessment in science education standards; standards for science content; science education program standards; and science education system standards. The message throughout the standards is clear: children learn best through scientific inquiry. "Inquiry into authentic questions generated from student experiences is the central strategy for teaching science."

The science teaching standards describe what teachers must understand and be able to do in the teaching of science.

- TEACHING STANDARD A: Teachers of science plan an inquiry-based science program for their students.
- TEACHING STANDARD B: Teachers of science guide and facilitate learning.
- TEACHING STANDARD C: Teachers of science engage in ongoing assessment of their teaching and of student learning.
- TEACHING STANDARD D: Teachers of science design and manage learning environments that provide students with the time, space, and resources needed for learning science.
- TEACHING STANDARD E: Teachers of science develop communities of science learners that reflect the intellectual rigor of scientific inquiry and the attitudes and social values conducive to science learning.

Standards such as these promote authenticity and community in the classroom. The autonomy that arises from true scientific inquiry provides the student with ownership vital for their positive experience of science and its processes.

Just as the science community has come to value inquiry, other disciplines have added an inquiry component to standards and benchmarks. The National Council of Teachers of Mathematics has written a set of curriculum and evaluation standards that encourage inquiry in math. In addition to viewing the learning of mathematics as a process in which children are to be actively engaged, the guide goes on to mention such verbs as *explore*, *justify*, *represent*, *solve*, *construct*, *discuss*, *use*, *investigate*, *describe*, *develop*, and *predict*. These processes "are used to convey this active physical and mental involvement of children in learning the content of the curriculum." All knowledge results from someone asking a question. Inquiry science promotes in children the habit of asking, seeking, revising their ideas, and asking again.

*Project 2061 consists of a series of recommendations designed to make Americans science-literate before the return of Haley's Comet in 2061.

Addressing Curriculum Concerns

"This is fifth grade. Kids shouldn't be using batteries. We covered electricity in fourth grade."

This is a logical concern, and although it sounds negative and judgmental, this comment probably grows from a misconception about curriculum design. A curriculum is created to make sure that all students receive instruction in important concepts. Therefore, certain ideas are addressed at each grade level. However, those ideas should be retained and utilized in succeeding years. Learning theory supports application as an important part of knowledge acquisition. Using batteries or mealworms in inquiry, even though they are not specified by the curriculum, is good educational practice. When such items are used before the grade level in which they are part of the curriculum, children develop prior knowledge, which will enhance their understanding. The best curricula are designed to include ideas—not to exclude them.

"You can't let kids decide what to do. They'll just goof off and do nothing."

There is comfort for many teachers in the notion that the curriculum, followed step by step, ensures that students are learning. Often, however, students are merely doing what they have been told to do, or following written steps without understanding the process. This leads to what is sometimes called "hands-on, minds-off" instruction. Actually, teachers need reassurance that it is okay to trust children. A child who is standing in front of the classroom library scanning for a title that will help in her research is engaging in complex thinking—deciding to research, knowing where to seek a text, and making a logical choice. This isn't "doing nothing"—this is learning.

"The school year is too busy to overlap topics from one year to another. Plus, the children will get bored."

Somehow, there arose a myth among educators that "once is enough." Electricity in fourth grade, rocks in fifth; repeat a topic and risk hearing the cry from students, "We've already *done* that!" However, the teachers who teach these kids plan their own gardens at home year after year, designing a layout, choosing plants, then adapting when the rabbits eat their impatiens, and they never get bored. The difference is in ownership.

When students pursue answers to questions, they may revisit science topics that were introduced in earlier years. The work they do is not scripted in a text; their procedures are based on what happens as they experiment.

Seven

The Language Connection

Inherent in science is communication. Science as a community cannot exist without it. Any discovery or observation that goes unreported is lost. Scientists communicate! Not only do scientists document their own findings, they also seek out the findings of others. What we do builds upon the documented successes and failures of ourselves and other scientists.

Encouraging children to do the things that scientists do means enabling them to communicate effectively. There is an immediate and authentic reason for children to speak and write about their experiences and for them to listen and read about the experiences of others. Indeed, language creates and makes possible the community we are building. The major goal of our schools is to help children communicate, and inquiry science provides the need and the desire among the children to do so.

Children's questions come from experiences, from observation, from information in books. Sometimes they are inspired by unlikely situations.

Sarah's Question

One day Sarah was reading *Hello, Mrs. Piggle-Wiggle* by Betty MacDonald. She was reading about Melody, a character in the story who cried often about almost anything. Melody's father said that she cried so much, he could rent her out to water lawns in the neighborhood with her tears. Cornell, Melody's brother, pointed out that the salt in the tears would kill the grass.

This is where Sarah stopped and began thinking about salt water and tears and the effect of salt on grass. What a great testable question. Does salt water really kill grass? Sarah immediately wrote her question on the question board and began thinking about ways to investigate it.

"We can try watering grass with salt water," Sarah said, "but how much salt should we use? How salty are tears? How salty is the ocean?" The questions began to flow.

Trying to assist the children in seeing questions all around them is an inquiry teacher's goal from the start. Devising questions from the books they are reading can be quite interesting for the students. Figure 7-1 shows a tool called the Science/Literature Connection. When introduced to the class, this form is first used during read-aloud, with questions modeled by the teacher. Once the children catch on, it becomes almost a game as they model questions for each other. For example, while reading aloud Laura Ingalls Wilder's *Little House in the Big Woods*, we learned that the children in the story each received a pair of mittens for Christmas. I stopped and asked the class about gloves and mittens. Which keep hands warmer? Does color matter? What about different materials? How could we find out? Here were some testable questions that were not asked by the book we were reading, but were inspired by reading it. My hope is that the children will find their own questions in the books they are reading and then perhaps go on to investigate them.

More with Top-Shelf Books

Top-shelf books, the collection of special science trade books that live on the top shelf in our classroom (see Chapter 2), can be used for more traditional language arts lessons and activities while still adding to the inquiry process. One of the problems I had when I first realized the value of trade books was that we only had single copies. It was difficult for small groups to read a book and then discuss its content when there was only one copy of the book in the room.

To solve this dilemma I have begun grouping some of the books from the top shelf along particular themes. Each book group contains about eight or nine books. Students pick a theme, share the associated books, and form small reading/discussion groups. After several days, the students have read every book.

Listed below are some of the themes, with the trade books used.

Earth Cycles:
The Sun (Branley)
How to Dig to the Other Side of the Earth (McNulty)
Sierra (Siebert)
Hill of Fire (Lewis)
Fossils Tell of Long Ago (Aliki)
Throwing Things Away (Pringle)
Inside the Earth (Cole)
Magic School Bus Inside the Earth (Cole)

Outer Space:
Space Words (Simon)
Journey to a Black Hole (Branley)
Galileo and the Universe (Parker)
Comets (Branley)
Planets in Our Solar System (Branley)
Shooting Stars (Branley)
Neptune (Branley)
Saturn (Simon)

Science/Literature Connection

Name ____________________________________
Book ____________________________________
Author ____________________________________
When Read ____________________________________

Testable Questions

Think of a testable question based upon the book you read.

____
____
____
____?

What do you think the answer might be? Why?

____
____
____
____
____
____

Design an experiment that might answer your question.

MATERIALS:

____
____
____

FIRST I WOULD:

____
____
____
____

NEXT I WOULD:

____
____
____
____

THEN I WOULD:

____
____
____
____

Extra: Complete your experiment.
Report your results in the BOOK OF DISCOVERIES.

FIGURE 7-1 *Science/Literature Connection*

Our Atmosphere:
Tornado Alert (Branley)
Getting Oxygen (Kramer)
Hurricane Watch (Branley)
What Will the Weather Be? (DeWitt)
Flash, Crash, Rumble, Roll (Branley)
Something Is Going to Happen (Zolotow)

Web of Life:
Snakes Are Hunters (Lauber)
How a Seed Grows (Jordan)
One Small Square (Silver)
Germs Make Me Sick (Berger)
What Happened to the Dinosaurs? (Branley)
How Do Apples Grow? (Maestro)
Look Out for Turtles (Berger)

People and Nature:
The Lady and the Spider (McNulty)
A Snake Mistake (Smith)
Bob and Shirley (Siefert)
Squirrel Watching (Schlein)
Oil Spill (Berger)
Bentley and Egg (Joyce)
The Puffins Are Back (Gibbons)

Scientific Thinking:
How to Think Like a Scientist (Kramer)
Townsend's Warbler (Fleischman)
How We Learned the Earth Is Round (Lauber)
Auks, Rocks, and the Odd Dinosaur (Thomson)
Always Wondering (Fisher)

Students are given the book group list of activities (Figure 7-2) from which to select their exercises. The number of activities they are asked to complete can vary. The Science Book Discussion Script (activity number 2) is shown in Figure 7-3. This activity has worked especially well in directing the students to conduct their own discussions based upon the books they have read. I select the student leaders either on a rotating basis or when strong leadership is needed. Although the discussions are student led, it is essential for the teacher to be visible and to monitor each group.

Top-shelf books are also excellent for student-produced book reviews. Not only are book reviews a way for the teacher to assess who is reading what books, but when bound in a class folder, the reviews become a resource for students to find titles they might want to read in their entirety. Finally, the book reviews are another reading response which fit nicely into a language program.

Ecological Mysteries

A genre of literature that incorporates inquiry science is the ecological mystery. In these books, characters struggle with a variety of questions and problems in order to solve mysteries related to the environment. The reader is in on the hunt for the solutions, considering clues and pieces of evidence uncovered by the characters. Eco-mysteries help reinforce the importance of questioning, gathering information as part of an investigation, and drawing conclusions in response to the data. Finding literature that supports the activities we are doing in the classroom helps to validate our own investigations.

Top-Shelf Book Groups

Name ____________________________

Activities: ____________________________

Book Group: ____________________________

Complete any ____ activities listed below with the books from your selected group of books.

____ 1. Select any two books from the group and complete a Venn diagram. Include at least three entries in each section.

____ 2. Participate in a group discussion using the Science Book Discussion Script.

Other students participating: ____________________________

Date ________________

____ 3. List one new question you thought of after reading each book. List the book titles and questions. Then, write your favorite question on the question board.

____ 4. Find a fact in any one book. List that fact and the book title. Next, find the *same* fact in another source. List the other source and its author.

____ 5. Make a poster for your favorite book from the group. Include the book title and author. (Your poster will be displayed for others to see.)

____ 6. Write a letter (in correct form) to any author of a book in your group. Later, your letter may be sent to the publisher for delivery to the author.

____ 7. List any books from the book group that are also in our school library. Include the book title, author, and the Dewey decimal number.

____ 8. List the books from the group in order from least favorite to favorite. Explain why your top-rated book is the one you like best.

____ 9. Make an advertisement for any one book in the group. Include the book title, author, and publisher. Also include a brief description telling why the book is useful and a picture. Later, we may show your advertisement to the librarian to try to convince him/her to order the book.

____ 10. For any one book, prepare an index of at least ten items. Your index should include an alphabetical listing of the topics with page numbers.

Your activities are due on ______________

FIGURE 7-2 *Sample Activities for Top-Shelf Book Groups*

Science Book Discussion Script

Directions: Meet with your group. The leader will use this script to direct the discussion.

LEADER: *Would someone please tell about one of the books read?*

ANY STUDENT: *(Tell about a book.)*

LEADER: *Would anyone like to add anything about the book?*

ANY STUDENT: *(Permit time for discussion.)*

LEADER: *Would anyone like to tell about another book read?*

ANY STUDENT: *(Tell about another book.)*

LEADER: *Would anyone like to add anything about the book?*

ANY STUDENT: *(Permit time for discussion.)*

LEADER: *Does anyone have a recall (fact) question to ask the group about one of the books?*

ANY STUDENT: *(A recall question is asked and discussed.)*

LEADER: *(Permit time for several more recall questions and discussion.)*

LEADER: *Would someone please ask the group a cause-and-effect question?*

ANY STUDENT: *(A cause-and-effect question is asked and discussed.)*

LEADER: *(Permit time for several more cause-and-effect questions and discussion.)*

LEADER: *Would someone please ask the group a compare-and-contrast question?*

ANY STUDENT: *(A compare-and-contrast question is asked and discussed.)*

LEADER: *(Permit time for several more compare & contrast questions and discussion.)*

LEADER: *Would anyone else like to ask the group any other questions from one of the books?*

ANY STUDENT: *(Any additional questions are asked and discussed.)*

LEADER: *Which book does the group rate as the best? Please explain why.*

ANY STUDENT: *(One of the books is identified as the favorite. Reasons are explained.)*

LEADER: *(Permit time for other books to be discussed as possible favorites.)*

LEADER: *What in the books we read might have to be explained for a second- or third-grade student? (Permit time for discussion.)*

LEADER: *Thank you for your participation!*

Note to Leader: Please list the names of your discussion group on the back of this paper. Circle the names of the students who contributed to the discussion.

FIGURE 7-3 *Science Book Discussion Sheet*

One especially interesting author of eco-mysteries for children is Jean Craighead George. In books like *Who Really Killed Cock Robin?* and *The Missing Gator of Gumbo Limbo*, Ms. George provides role models through her characters who show how scientific inquiry can help unravel ecological mysteries.

Early in the year I like to read aloud *The Fire Bug Connection*, which helps to set a tone for scientific inquiry. Characters in the book are puzzled by the inability of a particular species of firefly to mature and display its wonderful color. The children in my class are taken along on a journey of mystery and investigation. They easily identify with the child-scientists in the story as they share the excitement and frustration of the mystery and meet adult scientists in the book who assist through their own questioning. The characters are wonderful role models. The students witness how the characters think and, even more importantly, how they *rethink* situations when things don't work out.

Jean Craighead George includes numerous scientific facts in her stories for the readers to ponder. For example, twelve-year-old Maggie in *The Fire Bug Connection* loves to examine insects. One of Maggie's favorites is the bombardier beetle. This beetle brews enzymes and hydrogen peroxide in its abdomen for defense. When provoked, the aptly named bombardier beetle can shoot a 212-degree-F droplet of chemicals, which on one occasion brought an instant blister to Maggie's hand. These are the kinds of facts that children find intriguing and that lead to a wide array of questions. During read-aloud, the children are always welcome to go over to the question board to write down whatever questions may come to mind. It is fascinating to see students immediately respond to what is being read. From *The Fire Bug Connection* came such questions as:

- Do bombardier beetles live around our school?
  —Ryan
- How is a bombardier beetle able to hold acid and chemicals in its body?
  —Zach
- What other kinds of beetles are around our school?
  —Kim
- Is the acid in bombardier beetles there when they are young?
  —Thea
- How long do beetles live?
  —Nathan

These are relatively simple questions, yet they can lead to much more. First, they begin to establish a connection to past experiences. Further discussion reinforces their understanding of these experiences. Plus, some of the questions led to investigations around the school yard. Recess became more than just play; it was often an expedition bringing back more questions than answers. Utilizing literature to inspire questioning helps the children to find connections with science in all that they do.

From Reading to Writing

One of the toughest tasks for children to do is to write. When confronted with a blank sheet of paper, many children are frustrated because they feel they have nothing to write about.

So far, as science writers, the children have completed log pages in the discovery box folders. Questions, descriptions of investigations, sketches, and accumulated data have been recorded. The children have been making entries in their own dialogue journals. They have written about past experiences in science and have expressed their own questions on the question board. The children have been documenting their discoveries in the Book of Discoveries for others to read. All of these activities, although excellent ways to record thoughts and ideas, are only preliminary steps in a more formal writing process.

Often children are asked to select a topic with which they are familiar to write a story or article as part of a writer's workshop activity. Looking back on the journals and folders as prewriting tools can assist the students in organizing their ideas. It is important for them to see that what they wrote earlier was not merely an assignment, but may be authentically used later to assist in their further work. As their teacher, it is my role to encourage the reluctant writer by pointing to all the preliminary writing he or she has already done. The children themselves are experts in whatever field they may be investigating, and they are well suited to write about those topics. The toughest part, getting started, has already begun. Our challenge is to encourage them to effectively communicate what they have accomplished.

The traditional science research report provides practice in formally gathering facts from other authors. By reading and then reporting on what other scientists have discovered, the children become familiar with the topic. This experience is crucial. Scientists reflect upon and respond to what is already known. In their research reports, my students often cite references to past KIC journal authors. Writing research articles and citing past students strengthens our community of scientists.

Writing Trade Books

The writing children do for other children can take a different form. Using trade books as models, students write their own picture books to reflect what they have investigated and discovered. One example was *The Pyrite Who Thought He Was Gold*, written by Hannah, Julie, and Tiffany.

The authors had been learning about the characteristics of minerals from our science curriculum and their own investigations. They were fascinated that pyrite (fool's gold), although shiny and pretty, had relatively no value when compared with gold. The girls wrote a picture book about a proud piece of pyrite who thought he was gold. Some doubt was cast upon his true identity, so pyrite agreed to submit to a series of tests. The authors described the mineral tests they had been using during science. Of course, the crucial challenge was the hardness test, which proved pyrite not to be gold. In the end, pyrite was consoled by the other minerals, who assured him that they liked him just the way he was.

The pyrite book (its entirety is shown in *Science Workshop*) was a book full of facts portrayed in a fun way. The authors revealed an understanding of the importance of mineral testing for identification. They demonstrated that sometimes minerals are identified incorrectly and that "experts" (be they geologists or students) have the ability to find information through a systematic set of procedures. Student assessment of the understanding of these concepts was clear and easily made. Furthermore, in their writing of a real book to be read by others in the years to come, the authors were authentically motivated to be careful with spelling, punctuation, and mechanics.

Writing Their Own Ecological Mysteries

Once children enjoy reading a particular style or genre of book, they often like to write their own. Eco-mysteries can involve children in more complex writing.

"Writing ecological mysteries," explains Jean Craighead George in *Beyond the Science Kit*, "is really writing a scientific investigation. It's a translation of the scientific process into a form of literature called a mystery." Ms. George was speaking to a group of students about to begin their own eco-mysteries.

After exploring the school yard, the children identify problems or questions they observed. Answers or solutions are quite likely unknown. As the children investigate further (both through reading and making observations), answers or solutions may emerge. As Ms. George describes it, "Unlike scientists, you might know the answer before you begin your story, but as a good mystery writer, you save it until the end."

Mystery topics might include such questions as:

- Why are the fish dying in our tank?
- What kinds of seeds are these and where did they come from? How did they get here?
- Why are there so many (or so few) monarch butterflies in our school yard?
- Why are there so many (or so few) spiders around our building?
- What are these tracks in the snow? What was this animal doing so near our school? Is there a story in the tracks?
- What are these insects in the pond mud? Will they change into another form?
- Why didn't this science experiment turn out the way I expected?

Other questions might originate from data recorded on the Life at the Outdoor Classroom sheets, as described in Chapter 5. Questions and mysteries would be present if certain species were seen to be in decline. For example, having a baseline population of salamanders could highlight problems if their numbers were to decrease. A possible solution to the mystery might be a recent project to clear the forest of fallen trees and logs. The project may have beautified the forest and perhaps minimized fire hazard, but it would have also decreased the habitat for the salamanders. These questions and situations could lead to fascinating mystery stories, each based upon an observation, question, or problem. An extensive guide to writing ecological mysteries (even at school locations with minimal school yard facilities) is detailed in *Beyond the Science Kit*. Included are student samples and graphic organizers that help initiate the project.

Writing for Grants

One of the challenges of inquiry science in the classroom is finding the resources to secure the many materials needed to sustain ongoing investigations. By writing for grants, the children can take part in this process by persuading others that their projects deserve support.

There seems to be an endless number of grant possibilities, many available for specific purposes in local areas. Quite a few are particularly partial to requests written by students. For my student scientists, however, small, immediately available funds were needed. To meet this need, I designed an internal grant application.

I first realized that internal grants were needed when some students wanted to conduct research on electricity and motors. They needed large lantern batteries and several direct current motors that were not available in our science supply closet. Unable to personally fund the growing number of investigations, and having students whose parents were not always able to buy the materials needed, it soon became apparent that a source of funds was required to keep our inquiry going.

We convinced our PTA to set aside three hundred dollars for an inquiry grant program. The process has worked like this. Students who need materials for their investigations are invited to apply for a grant from the PTA inquiry grant program. The form (Figure 7-4) was developed based upon actual grant request forms. The children are asked to describe their research question, list the materials needed, and explain how the investigation will be conducted and evaluated. If the grant is approved by the grant committee, the student(s) are expected to furnish a report to the PTA upon the completion of the research. Grants awarded have been in the range of ten to twenty dollars each. Materials purchased have been batteries, wires, chemicals (vinegar, cornstarch), and simple building supplies (oak tag, special glues, and tapes).

The grant committee itself can take one of several forms. A committee of teachers and students might meet periodically to review proposals and distribute funds. Or, to keep the process simple, one or two teachers or PTA members might form the committee. Whatever the form of the committee, it is essential that a clear message be sent to the students: the grant proposal application must be completed clearly and in its entirety. Requests have been returned to applicants because of missing information or lack of details. The students realize that convincing others to grant them money requires effort. They also learn that modern science requires more than just *doing* science; scientists today must be able to find funding for their projects. Early sophistication in these areas will prove useful later. Moreover, business leaders repeatedly point out the need for students to learn clear communication skills, and the research grant is an authentic practice method.

Inquiry Grant Proposal Application

Manchester Elementary PTA

Manchester Elementary students who are engaged in or planning a scientific investigation are invited to apply for financial assistance to further their research. The PTA Inquiry Grant committee is interested in all areas of scientific inquiry. Please describe your investigation by completing the spaces below.

Names of students working together on this project ____________________

____

Teacher(s) ____________________ Grade level ____ Date __________

Describe the testable question that this research will attempt to answer.

____

____

____

____

BUDGET: List the materials needed, the quantities, approximate costs, and sources.

MATERIALS	QUANTITY	APPROXIMATE COST	SOURCE

TOTAL AMOUNT OF GRANT REQUEST $__________

Describe how your group plans to use the materials to answer the question. (Provide a step-by-step procedure.)

____

____

____

____

____

____

____

____

FIGURE 7-4 *Inquiry Grant Proposal Application*

Provide a schedule of your project. (Include start date, major milestones, and completion date.)

____

____

____

____

Describe your plans for evaluating the success of your investigation. (How will you know if you are successful?)

____

____

____

____

Has your group applied for or received financial support from other sources? If so, please describe.

____

____

____

____

If this grant request is approved, a written report will be required upon the completion of the project. The report is to describe the testable question, materials used for the investigation, how the investigation was conducted, and results of the investigation.

By signing this grant request, the students agree to the provisions described and indicate that the information contained in this application is accurate.

____

____

____

____

____

SIGNATURES OF STUDENT SCIENTISTS

A RESPONSE TO THIS GRANT REQUEST WILL BE PROVIDED WITHIN 2–3 WEEKS.

FIGURE 7-4 *(Continued)*

Eight
Additional Inquiry Enhancements

Initiating an ongoing inquiry approach to science requires much effort from the outset. Sustaining it requires constant monitoring. By midyear, students and teachers see the rewards of such an endeavor. The children are engaged, happy, asking questions, and reflecting upon the experiences and investigations of others. And in most cases, they want to do more.

Extensions of what was begun earlier help to keep the process fresh. New experiences keep the investigative fires burning. The potential for questions, investigations, and further discoveries seems to lie within all that we do. By this time, the children have no doubt that they themselves are scientists. They show more perseverance and a greater degree of ownership in their investigations. As a class, they are becoming a community that is dynamic and supportive of its members and their common goals, yet questioning of one another in positive ways. Each class may not progress in the same way, but I am always amazed by the strides that are made.

Moving Beyond Discovery Boxes

By midyear, many students express interest in a wider range of investigations. Their own imaginations have pushed them into areas that require more or different materials, or topics not accessible in the available discovery boxes. Many have pointed out that one hour with a box is not nearly enough time to accomplish the goals they are setting for themselves. The mealworm investigations, for example, require setup and observation that takes days or weeks. Students engaged in taking items apart or working with the ramps are spending time daily with their materials and experiments. As a result, the inquiry periods begin to change. Rather than scheduling periods in which the students formally sign up for and use particular discovery boxes as described in Chapter 4, time is made available for students to work on a variety of projects. Some may continue with ongoing investigations that may be extensions of the discovery boxes, while others may work on projects inspired by completely different experiences. The discovery boxes are not put away, but for some students their use is of a more independent and far-ranging nature.

Some students may never reach the point of independence at which they (and their teacher) feel comfortable moving beyond the structured use of the boxes. For those students, the boxes are still a meaningful way to engage in inquiry. Others, however, become so involved in their own investigation and desire to do more that they will be engaged with their pursuits during the entire day if permitted. For those students, contracts offer a way to reserve time for their growing obsession with scientific discovery.

Inquiry Contracts

Contracts, of course, have long been used by classroom teachers to individualize instruction. Basically, a contract is a promise, usually made in writing. Two or more people agree to do or provide something for each other. A contract shows who has made the agreement, what each person promises to do, and when they promise to do it. Parties engaged in the contract sign the document, and each get a copy for their own records.

Contracts play a vital role in any classroom in which the students are autonomously engaged in inquiry. They clearly define the expectations of both the teacher and the student. A contract provides accountability in an environment of divergent activities. Contracts in our class contain two promises by the teacher: first, to provide *time* for the student to work on whatever requirements the contract contains, and second, to provide *exemptions* from selected assignments on which the rest of the class will be working. The student promises to fulfill whatever requirements are described in the contract by the due date shown. Each contract has a reading and a writing component, and something the student is expected to do. Exempting students from selected assignments is a great way to encourage them to obtain and sign contracts. Virtually all of the time, contracts will require more effort in meaningful pursuits than a regular class assignment might demand. However, the choice and freedom of selecting a contract and independently embarking on an investigation is compelling to most students.

Figure 8-1 shows a basic contract, which provides time for the student to work on an investigation using a discovery box. The contract is given a number by the teacher for record-keeping purposes. A reasonable due date is also selected. The student identifies the discovery box to be used, poses a testable question, and is required to read a negotiated number of books on the topic (usually three). In our class, students copy the correct bibliography form into their language arts journals early in the year. The journal is to be consulted for that information.

The contract also directs the student to prepare a contract journal. These are small booklets with construction paper covers. Lined pages are stapled within to

Teacher/Student Contract

Contract Number ______ Date __________________

______________________ agrees to work on the following long-range assignment:

USE A DISCOVERY BOX FOR AN INVESTIGATION

to be completed on or before __________________________

The classroom teacher agrees to provide a reasonable amount of class time by exempting the student from selected classroom assignments. Completion of this contract (will/might/will not) require additional time at home. The following provisions apply in the completion of this contract:

- A contract journal will be prepared by the student.
- The student plans to use the _________ discovery box.
- The student will begin with a testable question, which the materials in the discovery box will help answer. The question is to be written in the contract journal.
- During the investigation, entries will be made in the contract journal. Details should be sufficient for others to repeat the activities. Discoveries will be added to the Book of Discoveries.
- Prior to the contract expiration date, _________ books on the topic will be read and added to the contract journal in correct bibliography form (as shown on journal page _________).

The student agrees to do his/her best work on the completion of this contract.

____
STUDENT

____
PARENT

____
TEACHER

______
CLASS

FIGURE 8-1 *Teacher/Student Contract for Using a Discovery Box*

form a tablet. The students prepare the journal by writing their name, contract name, and contract number on the cover. Contract journals are convenient because each is dedicated to a particular contract. All notes and records, diagrams, and measurements that are accumulated while working with a contract are placed in this journal. When the contract is completed, the journal is a permanent record to be turned in to the teacher. Moreover, the contract journal is also an excellent prewriting tool for a more formal written article or for preparing a presentation.

Inquiry-related contracts in our class include such titles as:

- *Invention Workshop*. Students read about past inventions and are encouraged to develop an invention of their own.
- *Diet for Mealworms*. Students investigate foods preferred by mealworms.
- *Mystery Seeds*. Unknown seeds are examined and a means of identifying them is developed.
- *UV Bead Research*. Independent research in the detection of ultraviolet rays as described in *Inquiry in Action* section in Chapter 2.
- *Write a Science Trade Book*. Opportunity is given for students to write their own science books as described in Chapter 7.
- *Conduct an Inquiry Investigation*. Open-ended opportunity for students to engage in their own investigations.

For the Conduct an Inquiry Investigation contract, students are required to complete an Inquiry Investigation Plan as shown in Figure 8-2. This form helps the student to map out a plan for the investigation by considering materials needed, steps to be taken during the investigation, and an estimate of time requirements. It is essential when negotiating a student-selected contract that the teacher furnish the structure for clear expectations and documentation before, during, and after the investigation.

One student's contract last year was titled *Grow a Vegetable*. Not only did it provide food for thought, but it also provided a snack for the class. Emily had read and signed the contract that required her to plant a vegetable seed of her choice and observe its growth. She chose to grow a cucumber under the growlights in our room. In her contract journal, Emily wrote information from her research: facts about the origin, nutritional value, and uses of cucumbers. When the beautiful yellow flowers emerged, Emily and her friends used a paint brush to pollinate them. Then, after a few weeks, little cucumbers began to grow. Before long we were slicing the vegetable as Emily shared with the class the long process it was for them to grow. Everyone agreed that the little cucumbers tasted delicious.

Emily's plant wasn't finished, however. Its vine climbed ever higher on our rack holding the lights. One Monday morning we noticed that the tendrils of the cucumber plant had grabbed a cardboard bird on a bird mobile hanging near the lights. The class remarked and laughed that although they had heard of birds eating vegetables,

Inquiry Investigation Plan

Name(s) ______________________________

Date ____________

We are interested in ______________________________
TOPIC

We would like to attempt to answer the following testable question. ____________

____

We predict that ______________________________

____

The following materials will be needed for our investigation.

MATERIAL	SOURCE	MATERIAL	SOURCE

To answer our question, we will do the following:

(First) ______________________________

____

(Next) ______________________________

____

(Then) ______________________________

____

____

____

Sketch (with labels)

We will need _____ school hours per week, and _____ home hours per week for our investigation. Our investigation will take approximately _____ weeks.

Teacher Notes: ______________________________

____

____

____

____

____

STUDENT SIGNATURES

FIGURE 8-2 *Inquiry Investigation Plan*

this was the first time that they had heard of a vegetable capturing a bird. Emily was asked to please control her cucumber!

Visiting Scientists

Any teacher who values scientific thinking will welcome adult scientists into the classroom. As role models, adult scientists show the children how scientists behave and respond to questions and problems around them. Scientists have wonderful stories about their own research and investigations. It doesn't take long after a scientist enters the classroom for a rapport to develop; the children and the adult have so much in common as fellow scientists.

Scientists in classrooms today are playing far different roles than they did a generation ago. At that time, a scientist might come into a classroom for an hour or so, impart some interesting facts, dazzle the children with scientific pops and bangs, and then leave. The children would applaud and be amazed and then go back to their textbooks. Today that is all changing!

More and more scientists are expressing interest in engaging children in research, in involving the children in authentic inquiry investigations. A scientist might come into the classroom, discuss his or her area of expertise and research, give some background information, and discuss some equipment, and then invite the children to participate in the research. The teacher's effort to involve the children in real science is reinforced by the visiting scientist's acceptance of the children as scientists and collaborators.

In our class, children prepare all year for visits by scientists. They know what scientists do (remember the lists they made, as described in Chapter 2) and by now they are aware of the ways scientists think. After all, they have been doing a lot of science themselves. Before the first visit, the children and I review some of the vocabulary that may be encountered and share with one another what we already know about the topic.

Preparing the scientist may not be as easy. Many who are new to the classroom are quite nervous about standing before a group of children. One scientist told me at a planning meeting for classroom visits that "going to schools is more stressful than making a business presentation." The teacher's role is to reassure our visitors that although the children may be painfully honest, they will idolize anyone who comes to their classroom to share stories and who will value them as scientists.

One scientist who visited our classroom was Dr. Marty Condon, who was at the time with the Smithsonian Institution. Her visits with the students engaged them in authentic research that led to astounding results. Dr. Marty, an entomologist working with insects in Venezuela, came to our class and told the story of her research in the rain forest into flies that laid their eggs on a particular plant. What she knew was fascinating, but what she *didn't* know was what captivated the children. She needed help in identifying flies by examining patterns on their wings. Dr. Marty told the class that she suspected the flies were of differing species, but she had no way of differentiating the species in the field. After telling her story and explaining the problems she was having with her research, Dr. Marty asked the class if any students were interested in helping with her investigation. About a third of the class was quite interested, got contracts to work with Dr. Marty, and met with her on several occasions in the months that followed. The story of Dr. Marty, the students she worked with, and their investigation and discovery is told in the *Inquiry In Action* section at the end of this chapter.

Some large corporations supply scientists for classroom visits. They, too, are involving children in authentic research. Becton Dickinson is a biomedical manufacturing corporation with nearly twenty thousand employees worldwide. The Scientists On Call program through Becton Dickinson's Baltimore, Maryland, divisions provides to schools BD scientists who are committed to enhancing scientific curricula and promoting inquiry-based science. The scientists visit classrooms and engage the children in authentic research problems or challenges. Leaving behind materials, tools, and equipment, the scientists return several weeks after their first visit and review results of the student investigations that have been occurring in the classroom.

One year we were working with bacteria cultures and indicators. Jay Sinha and Mark Sussman of Becton Dickinson had visited the students. In their initial demonstration, they had shown how bacteria growth often raises the acidity (lowers the pH) of the material in which the bacteria are growing. They showed the students indicators that change color due to increased acid and are used in laboratories to detect the increase of the bacteria. Kathleen was one of the students experimenting with and observing the bacteria cultures. She used a blood plate, provided by Becton Dickinson. Using a special tool, she streaked the plate with sour milk using the four-quad method she had learned from Mark and Jay. Her question was: "Can I detect the presence of bacteria by streaking the milk on a plate?" Carefully watching and recording the growth of the bacteria colonies, Kathleen one day noticed something other than bacteria growing on the plate. She came to me and said, "I think there is a fungus on my plate."

"What makes you think that?" I asked.

"Well, I saw a picture of fungus and this stuff looks like the picture." She kept watching the colonies of bacteria and the new invader. Then Kathleen made a remarkable observation. The fungus, she said, was affecting the growth of the bacteria. The bacteria were dying as the fungus grew. When the scientists from Becton Dickinson returned, they were amazed with what Kathleen had noticed. They pointed out that this was quite similar to how penicillin had been discovered. Unknowingly, Kathleen had followed in the footsteps of Alexander Fleming, who had made a similar observation seventy years before.

Becton Dickinson has been generous with its contributions to inquiry science in our classroom and in others. In addition to supplying incubators and other kinds of equipment, BD has initiated a grant program that supplies funds to participating teachers and funds to students engaged in their own inquiry investigations.

Parent scientists, neighbors, and nearby corporations are all sources of visiting scientists. As an enhancement for inquiry, the scientists should have an authentic story to tell and then engage the children in a real investigation. Coming back to visit later communicates to the children that the adult values them as contributing scientists themselves.

Growlights

Bringing a part of the outdoors into any classroom can be accomplished with growlights. As an enhancement for inquiry, these are probably *the* one piece of equipment with the greatest effect. Plus, growlights are inexpensive. All that is needed is a shoplight fixture (about $10) and two growlight tubes (about $8 each). Suspend them from the ceiling over a table and you are ready to go.

When I first put up our growlights, I provided small flowerpots, topsoil, and a variety of seeds (purchased for ten cents a pack from a local hardware store). I told the children to go ahead and do with the materials whatever they thought would be interesting. Did they ever! They planted flowers and vegetables and carefully tended the seedlings as they grew. Each morning as the kids arrive, the growlight area becomes a gathering place where the children check on their plants and look at the plants of others. It is amazing how protective and careful the children become when their own seeds sprout and begin to grow. It doesn't take long for the children to realize the importance of watering. If they ignore their plant, it will die. They also learn about overwatering. Soon the plants need support. Then they flower and many produce some sort of vegetable. Wow! If ever there were an ideal tool for enhancing inquiry, the growlights must be it. Contracts are available for plant growth (like the one Emily negotiated to grow a vegetable). With seeds and plants come all kinds of questions and the possibilities for student investigations. Students experiment with colored light (using a variety of colored plastic), compare the effects of salt water and fresh water on plant growth, and try growing plants inside sealed terrariums with a variety of success. To do all this, students record and interpret data and observe carefully. If a teacher can select only one piece of new equipment, a growlight must be it. The possibilities are endless!

Animals in the Classroom

Soon after the start of each school year, the children ask about a class pet. I try to discourage the idea of a pet in our classroom, pointing out the many long hours we are away. I realize, however, how much students can learn by observing animals in the classroom.

In the fall, the many spiders found around the school are a wonderful way to begin our experiences with animals. As the spiders are captured, we discuss the ethics of keeping animals in our classroom. No animal is to be experimented with. They are guests in our room for observation only. Their safety and comfort must be considered at all times.

The National Science Teachers Association has published a Code of Practice (*The Science Teacher*, January 1986) that recommends acceptable uses of animals in the classroom. When students ask to bring in animals to keep in our classroom for an extended period of time, I ask them to complete the Animals in the Classroom form (shown in Figure 8-3). With this form, there is no question about what must be done to properly care for the animal and who is responsible for its care.

We have welcomed into our classroom a wide range of creatures for observation

Animals in the Classroom

Name ________________________

Several students have asked about the possibility of bringing animals into our classroom to stay for several days or weeks. If you would like to do so, please complete this form.

1. What kind of animal would you like to bring in? ________________________
2. Do you have a cage or container for your animal? ________________________
3. What must be done *daily* to care for your animal? ________________________
   ____
   ____
   ____
   ____
4. Do you promise to do these things *daily?* ________________________
5. What must be done *weekly* to care for your animal? ________________________
   ____
   ____
   ____
   ____
6. Do you promise to do these things *weekly?* ________________________
7. What food will your animal be eating? ________________________
8. Who will provide the food? ________________________
9. What other things do you know about your animal that might be important for its safety in our classroom? ________________________
   ____
   ____
   ____
   ____

Please read and sign the following agreement.

I have carefully considered the responsibilities of caring for an animal in our classroom and promise to do all that is necessary to ensure its comfort and safety. I further agree to care for the animal on a regular basis. Care includes feeding, providing water, cleaning of cage or container, and any other tasks as directed by the teacher. I understand that failure to abide by this agreement will require removal of the animal from the classroom.

____
STUDENT

____
TEACHER

PLEASE HAVE A PARENT SIGN HERE ________________________

FIGURE 8-3 *Animals in the Classroom*

and study, from spiders and insects to mice and hamsters. Once the animals are in our classroom, a multitude of questions follow. What do they eat? When do they sleep? How do they grow and change? Sometimes students ask about how animals feel or what they like. It is important for the children to realize that we can only guess about an animal's feelings or likes and dislikes. We can only observe behavior and then perhaps speculate based upon what we see.

Creating an animal's habitat can be an exciting project for the children. Providing a safe environment requires research into how an animal lives and what conditions it may need to survive. The Create A Habitat contract (Figure 8-4) has been used by students to build enclosures that are as close as possible to an animal's home in the wild. It requires initial research and then careful observation to assure safety.

Sources of animals are numerous. The sweep nets discussed earlier will supply a wide range of insects and spiders. Pond mud has been dredged with a shovel and placed in small aquariums. One spring we had an uncovered aquarium with pond mud containing the nymphs of a variety of insects. I will never forget when the class, working quietly on a standardized test, was suddenly startled by a huge dragonfly that had metamorphosed and emerged from the mud. It buzzed around the room several times before lighting on a wall for later capture.

Another source is the local pet shop. A variety of animals including mealworms and mice can be purchased at most shops at reasonable costs. Feeder mice cost only a dollar or two and are usually sold as dinner for snakes or other animals that require live food. Feeder mice in our classroom have a far brighter future. Sometimes, however, these little critters come with surprises.

One week, Earl brought in two mice that he wanted to observe. He had purchased them and their accessories with an inquiry grant from the PTA and had several questions he wanted to answer. He initially thought that the mice were both males. One, however, was pregnant and soon gave birth to about a dozen babies. Earl removed the other adult to ensure the safety of the babies and then went about observing and recording the behaviors of the mother and babies over the course of several weeks. We all learned quite a bit about the development of mice during those weeks and how hard the mother worked for their survival. The ultimate question soon became: Where would the mice go when they were grown? One of the students had a need for feeder mice for a snake. Earl was opposed to this, because the mice had become so familiar to the class. Other students wanted to take one or two home as pets but were soon dissuaded by their parents. Animals native to an area may be released in a nearby habitat. For example, crayfish captured in a nearby stream and brought in by the children may be returned to the wild. However, crayfish ordered from a supply catalog may quite likely threaten local crayfish species. These animals must be destroyed or donated to a local pet shop, which can dispose of the animals properly. Earl's mice were returned to the pet shop.

Animals in our classroom also include those that come to us. The fall always brings yellow jackets into our classroom, which get constant attention. Where they fly and when and where they land is always watched carefully by the children. Yellow jackets

Teacher/Student Contract

Contract Number ______ Date ____________________

________________________ agrees to work on the following long-range assignment:

CREATE A HABITAT

to be completed on or before ____________________________

The classroom teacher agrees to provide a reasonable amount of class time by exempting the student from selected classroom assignments. Completion of this contract (will/might/will not) require additional time at home. The following provisions apply in the completion of this contract:

- The student will prepare a contract journal.
- One or more books will be used for this contract for the student to learn about __________. Where this animal lives, what it eats, and how it grows will be researched. The books will be listed in the contract journal in correct bibliography form as shown on journal page __________.
- Based upon what is learned, the student will create a habitat for the animal. The habitat will provide space, food, water, and the proper conditions for temporary observation.
- A drawing or map will be made in the contract journal. Labels and measurements will be included.
- Daily entries will be written in the contract journal which describe the animal; its behavior, eating habits, health, etc.
- A short report (3–5 paragraphs) will be written in the contract journal about the animal based upon what the student read and what was observed.

The student agrees to do his/her best work on the completion of this contract.

____

STUDENT

____

PARENT

____

TEACHER

____

CLASS

FIGURE 8-4 *Teacher/Student Contract for Create a Habitat*

are interesting predators. One day we actually watched one pounce on an unsuspecting fly and remove its wings. Then the yellow jacket flew off, carrying the fly with it.

Animal guests, invited or not, will provide many experiences for the children. Observing, documenting, and predicting their behaviors greatly enhances the inquiry in our classroom. Our approach—that animals are guests rather than pets—encourages students to watch scientifically, inferring from their behavior what the creatures like, but never forgetting that these are wild visitors, not cartoon creations.

Models

Models in science serve several purposes. First, they permit examination and manipulation of items and concepts too large for practical scrutiny. A globe as a model of the earth enables the study of how the earth turns and how light and shadows are cast upon it by the sun.

Second, models can speed up time. A phenomenon that may take hundreds or thousands of years can occur in only a few minutes with a model. Continental drift is an example of an extremely slow process that can be more readily seen and understood with the use of a model.

Third, models can offer a safe means of observing an otherwise unsafe event. An oil spill in a model ocean can simulate the effects of such a tragedy without subjecting the environment to the actual event.

Finally, models can save money. Expensive scenarios can be played out with relatively inexpensive models to determine the possible effects of a phenomenon or event. Studies with earthquakes can thus be carried out.

Models can therefore be quite useful as students plan and carry out inquiry investigations.

The Model Mountain was devised for our classroom to study the effects of water erosion upon a mountain. Our science curriculum directed the teacher to construct a mound of earth outside and water it with a sprinkler or watering can. This activity made a wonderful demonstration, but did not lend itself well to independent study by the students.

The mountain (as shown in Figure 8-5) is a tub filled with playground sand. Suspended above the sand on a pole is a bucket in which numerous small holes have been drilled in the bottom. Water is poured through the bucket and falls as rain onto the mountain of sand. It has been used by students to observe erosion and then develop means of protecting land from the effects of rainfall. Following an initial demonstration of a rainstorm, students sign up for the mountain for a day or two to rebuild the mountain and conduct their own investigations. Before and after rainstorms, popsicle sticks, needlepoint webbing, skewers, toothpicks, and other materials are experimented with as means of erosion control.

As a model, the mountain speeds up time. An actual mountain would not erode as quickly as our model does. It enables the students to observe and try to remedy the effects of a much slower yet equally destructive phenomenon.

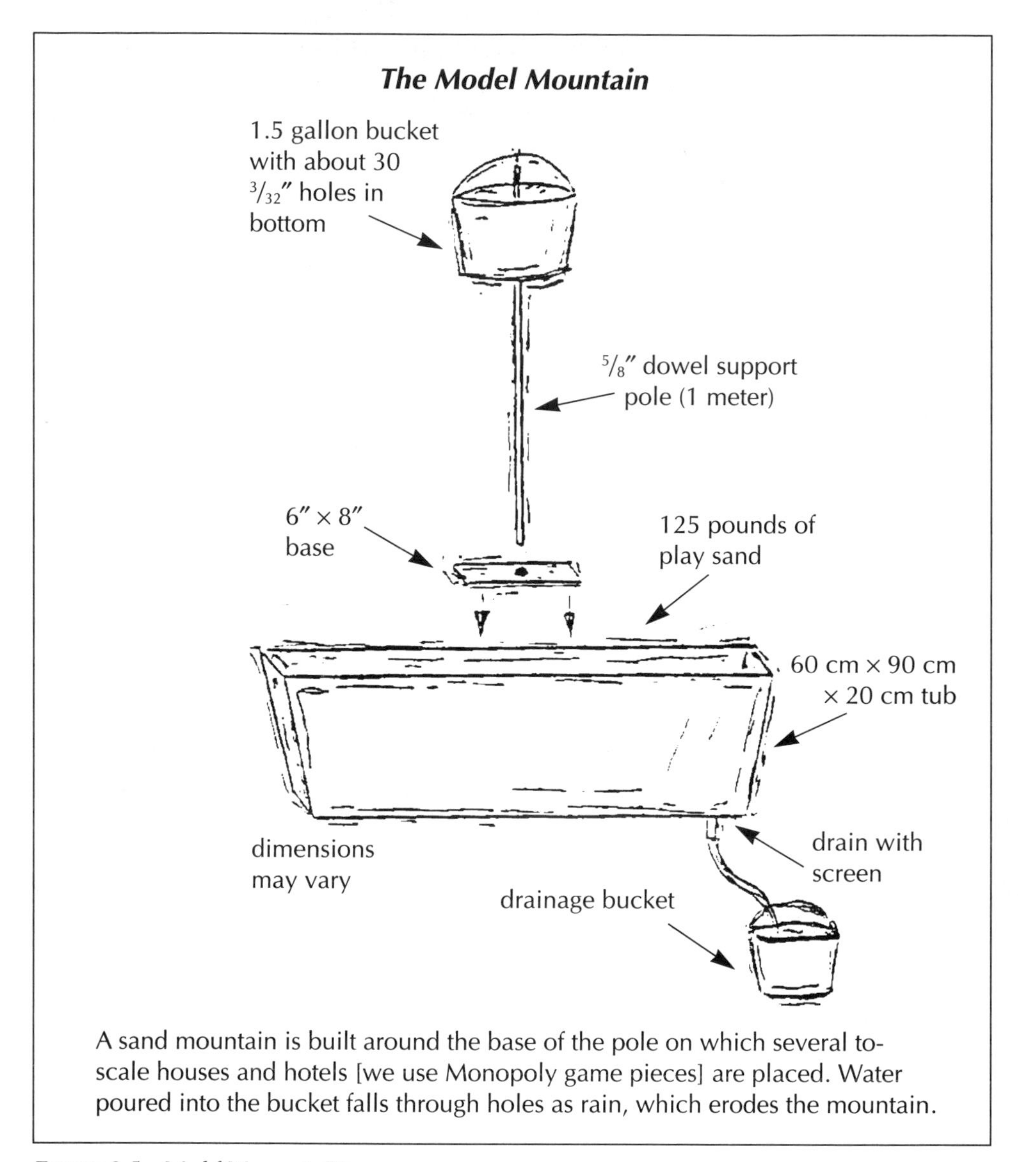

FIGURE 8-5 *Model Mountain Diagram*

Students keep records of their investigations. Figure 8-6 is a sample form from our model mountain logbook, which is added to by each team of students working with the mountain. As an ongoing record, the logbook allows students to review what others have tried before. The investigations and discoveries have led to some rather interesting explanations of water damage and control, the undermining effects of a high water-table, and the use of various devices tested on our model designed to conserve the soil. Since documentation is required, a record becomes available for other students to review. The logbook is also useful for assessment.

The Earthquake Machine is a model that can simulate a dangerous and costly event in the safety and controlled conditions of our classroom. Earthquake-simulator models can be as simple as a gently rocking table, or as complex as an expensive machine purchased from a catalog. Our homemade earthquake simulator consists of a twelve-by-eighteen-inch plywood board, one-quarter-inch thick, mounted on four one-and-five-eighth-inch springs. (I use Serv-A-Lite #91 springs. These or similar springs are available at most hardware stores.) The board can be manually rocked to create an earthquake. I have also added a small battery-powered motor to the underside of the board. The motor is attached to a weighted arm, which creates a circular motion to the board when turned on. The effect is quite similar to an actual earthquake. A variety of materials are used to construct buildings and towers on the platform of the machine. We have used base ten-blocks (counting blocks from the math lab) and have also tried building tall, swaying towers out of toothpicks and marshmallows. This easily constructed device has added an important inquiry component to the earth science portion of our science curriculum.

Model rivers and streams can be constructed with kitty litter pans and diatomaceous earth as described in *River Cutters* (Kaufmann 1989). Miniature rivers form as colored water trickles into the moistened material from a suspended plastic cup containing a small hole at the bottom. After watching a river form and drain into a growing lake at the bottom of the tilted tray, students go on to build dams along the model river. Tiny mountain ranges through which passes are cut by the rivers demonstrate the power of moving water. We have experimented with burying in unknown locations the ends of Q-tips saturated with dried food coloring. As described in *River Cutters*, these form toxic landfills, and the only clue to their presence is when a nearby river begins to flow with a strange color. The students then use tiny sponges, pipettes, and little plastic dams to try to keep the "toxic" material from polluting the lake at the bottom of the tub. This is a difficult and often futile effort that demonstrates the importance of careful waste management.

Inquiry on the Internet

Available to children and teachers are hundreds of Web sites on the Internet that are ideal for enhancing inquiry. Listed on page 95 are just a few of the best. As with books, curricula, and teacher-directed activities, these Web sites should serve as

The Model Mountain

The Model Mountain is available to observe erosion and to experiment with methods of soil conversion. Please complete the spaces below before, during, and after using the mountain.

NAME OF BUILDER(S) ______________________ DATE __________

Make a sketch of the mountain before a rainstorm. Include any houses, erosion control structures (fences, supports, etc.) with measurements and labels. Views from more than one direction may be useful.	Make a sketch of the mountain after the rainstorm. Include any houses, erosion control structures (fences, supports, etc.) with measurements and labels. Changes should be particularly noted.

What have you discovered that may be useful for another time? ______________________

____

____

NOTE: *Your records are important for others. Please complete additional pages for other mountain trials.*

FIGURE 8-6 *Model Mountain Data Sheet*

motivators for inquiry, not replacements for student-initiated investigations. And as is increasingly common with computer simulations, nothing should take the place of student participation with actual materials that can be handled and manipulated to yield authentic data. These Web sites should help in providing ideas for genuine inquiry.

www.nasa.gov (National Aeronautics and Space Administration [NASA])
Contains scientific and technical information, launch schedules, educational resources, and great links.

www.noaa.gov (National Oceanic and Atmospheric Administration [NOAA])
This site links to weather information, the National Weather Service, and up-to-date oceanic and atmospheric research.

www.nhc.noaa.gov (National Hurricane Center)
Here you will find information, forecasts, satellite-visible imagery, reconnaissance-aircraft data, and a hurricane tracking chart.

www.ideo.columbia.edu/EV/Earth View Home.html (Map Blast!)
Map Blast! Allows students to access a map for any location. Easy to use.

www.nssl.noaa.gov (National Severe Storms Laboratory)
This site, dedicated to information about storms, includes scientific research and a kid-friendly "weather room" with games and more.

www.telescope.org/rti (Net Telescope)
Visitors to this site can actually control a robotic telescope in the United Kingdom. After registering and gaining a password, a visitor can aim the telescope at any object in the northern sky.

volcano.und.nodak.edu (Volcano World)
Sponsored by NASA, this site contains information about volcanoes. Especially interesting is the opportunity for Internet conferencing with a volcano expert.

nyelabs.kcts.org (Bill Nye the Science Guy)
Nye Labs offers tons of easy demonstrations that could serve as a springboard to inquiry. In addition, this Web site contains full program listings for the popular series.

www.weather.com (The Weather Channel)
What a Web site! Besides current forecasts for cities across the United States, this site also provides weather maps, temperature information, and articles about weather events.

www.exploratorium.edu (The Exploratorium)
This Web site contains information and activities for both students and teachers. Electronic exhibits, interactive demonstrations, and links to current sites of interest are updated frequently.

www.nwf.org (The National Wildlife Federation)
On the main page, teachers can learn about environmental issues. On the kids' page, students can play games and read information at their own level. This site

also contains a terrific section on school-yard habitats—planning, implementing, funding, articles, resources, and more. For environmental information, this site has it all!

www.aqua.org (National Aquarium)
What goes on behind the scenes at the National Aquarium in Baltimore? How can a student set up his or her own aquarium at home? Students can find the answers to these questions at this Web site, as well as learning more about the animals who make their homes in the tanks.

www.seaworld.org (Sea World)
Information abounds at this site, which includes specific animal facts as well as career opportunities. Students can even call a toll-free number to obtain information.

www.nceet.snre.umich.edu (Environmental Education Link)
EE-Link is designed for teachers and others who need to obtain environmental education information. This site includes links to major organizations and projects and a calendar of events.

www.sierraclub.org (Sierra Club)
This site also supplies many environmental ideas and links.

www.usgs.gov (U.S. Geological Survey)
This searchable site contains earthquake information as well as climatic data and other items of interest.

www.acorn-group.com
Where do you find sweep nets? Visit this Web site to find the supplies you need to make inquiry work—sweep nets, field guides, magnifying glasses, and other odds and ends. Also find information on owl pellets and bone-identification charts.

www.fs.fed.us (USDA Forest Service)
News, products, services, and other related information. Includes information about owl pellets, wildlife, and more.

www.enc.org (Eisenhower National Clearinghouse for Mathematics and Science Education)
A database of K–12 programs for teaching science and math including ideas on integrating math and science with literature.

www.fi.edu (The Franklin Institute)
This site takes you to the world famous Franklin Institute science museum in Philadelphia.

www.garden.org (National Gardening Association)
Information about growing plants indoors, gardening grants, and more.

www.MonarchWatch.org (Monarch Watch)
This site is dedicated to the preservation of Monarch butterflies. Contains many classroom ideas for raising Monarchs, growing milkweed plants, involving students in fall-migration data gathering, and more.

eelink.umich.edu
This is the place for environmental education links. Numerous activities on recycling, wildlife, air quality, and other environmental issues. Includes opportunities for networking with other classes.

Inquiry Is All Around

Sources for inquiry investigations seem to appear all around us. Whatever the source, the common elements are student questions, investigation plans, documentation of discoveries, and sharing with others. Children at virtually all age levels are able to participate in some form of inquiry science, addressing each of the elements of an inquiry approach.

Inquiry in Action: The Manchester Fly

"The fascinating part of science," Dr. Marty told the class, "is how each solution brings with it more questions to answer and more problems to solve." For several months, students in our class had been working with Dr. Marty Condon, an entomologist with the Smithsonian Institution. Together, they were combing through data gathered in a South American rain forest. Speaking to our class and showing her slides, Dr. Marty told about her research in Venezuela with fruit flies that lay their eggs on a particular rain forest plant. She explained the problems and mysteries of her research and how she and other scientists were trying to solve those problems. But Dr. Marty had more than a story to tell, she had an invitation for the students to join her in her research. A group of enthusiastic fifth-grade students volunteered to participate in Dr. Marty's investigation.

Charis, one of the students in the group, described Dr. Marty's research in her journal. "Dr. Condon has found different types of flies. Some are seed flies and some are flower flies. The problem is that these flies look exactly the same." Dr. Marty had no way of differentiating the flies in the field. One possible approach: examine the spots on the wings of the flies and find distinguishing patterns.

Dr. Marty had piles of photographs showing the magnified fly wings. The complex pattern of spots on each appeared to be unique. Fellow scientists at the Smithsonian Institution could find no distinguishing patterns. Dr. Marty surmised and hoped that perhaps the unbiased eyes of children might uncover what others had missed.

The group of students worked collaboratively. They discussed and debated, tested each others' theories, and carefully examined the data together. "The group meets with Dr. Condon about once a month," wrote Jacob, another student researcher. "Between visits, we study enlarged photographs of the fly wings. With these pictures we look for patterns in the spots on the wing to help determine sex, origin of the fly (flower or seed birthplace) and/or species."

When Dr. Marty visited, the group told her about their theories and what they had accomplished. The students often met independently to continue their research. Dr. Marty offered encouragement and discussed what she had been doing since her last visit. As the months went by, a sense of respect and trust had grown among this small community of scientists. Furthermore, a sense of what science *really* is was growing. "I learned many things working with Dr. Marty," wrote Jacob. "I found what a scientist's life is like and how hard it is. I also learned that working scientifically with others takes much cooperation."

A clue to what Jacob and the others really learned about working together as scientists was found in the words of Dawn. "March 19—We met with Dr. Marty once again. Today we talked about Jacob's theory that didn't work. We didn't find any new theories today because we argued most of the time about Jacob's theory. We tested him on his own theory and he got five wrong out of nine." Undaunted, Jacob continued to study the data and to try to convince the group of the validity of his theory and said that the process was frustrating but not discouraging.

The perseverance of the fly group brought about a result none had expected. As Dr. Marty reviewed the student theories and results from the many hours of work, she noticed a discovery made by the group that led her to the detection of a previously unidentified species of fruit fly. Not only had the students assisted Dr. Marty in her research, they were rewarded for their discovery by having the new fly species named after their school.

"It's the only insect named after an elementary school," Dr. Marty proudly announced a year later. *Blepharoneura manchesteri* has been identified as a result of the students at Manchester Elementary." And indeed, *B. manchesteri* was officially described in the 1994 edition of *Systematic Entomology*.

These young scientists, who added to our scientific body of knowledge, did not set out to make a historic discovery. They merely accepted the challenge of assisting Dr. Marty in a search for patterns hidden on fly wings. These students had no way of knowing that the long hours spent working together would culminate in an insect being named for their school. But as scientists, they had no doubt about what they were doing and why.

This story is not being told to enlist other teachers and students to seek new species of life. The likelihood of another such discovery being made this way may be remote. And certainly, the fly students' success was assured whether a species had been identified or not. The point is that, with student inquiry, we never really know what the ultimate results may be. Unlimited vision really can bring limitless results. As teachers, our role is to provide the kinds of opportunities that may yield surprises beyond our wildest dreams.

Nine

The Kids' Inquiry Conference

The mood on the bus that bright March day in 1993 was unusually subdued for a field trip. But then, this was to be a trip like no other. As I walked down the aisle handing out name tags, I could sense the excitement and anticipation and feel the nervousness in the air. Along with the children and parents were containers of vinegar, limestone chips, marine fossils from four hundred million years ago, radish plants, water, oil, and mealworms comfortably traveling in bowls of oatmeal. After months of planning and working, we were finally on our way to the first ever Kids' Inquiry Conference.

As the bus bounced along, I noticed the children looking over note cards together, organizing and checking their overhead transparencies and handouts, and double-checking their bags of supplies. Some were scheduled to give presentations about the research they had been conducting. Others were to provide hands-on experiences for the nearly one hundred students who would be converging on the university. Although I had confidence in my students, I likewise felt a pang of anxiety. Together we were venturing into new territory.

The previous months had been exciting. It seemed so long ago when we began asking those first testable questions. The children had become increasingly independent as they conducted their own investigations and made their own discoveries. They had scrutinized each other's work, had written articles for one another in which they had documented their research, made entries in the Book of Discoveries, and had presented the results of their investigations to students in other classes. Now it was time to share with the world what had occurred in the classroom.

Inquiry science is, by definition, authentic science. In our classrooms we create environments in which children ask real questions, delve into research for genuine reasons, and make discoveries that are their own. The excitement of Archimedes who, having made his famous bath tub discovery, went running through the streets to tell the world, can be seen in the faces of young scientists who have their own discoveries to share. The children on the bus that spring morning traveling to the first

Kids' Inquiry Conference were following in the footsteps of scientists like Archimedes.

The Kids' Inquiry Conference evolved from a desire to extend the results of inquiry to others beyond the classroom walls. I had been considering my own questions about science and how we teach it. Would it be possible, I asked, to fashion a conference for children similar to an adult science convention; a place where kids from different places could come together for the purpose of sharing with one another the stories and results of their own scientific investigation? My hope was to provide opportunities for children to

- share the excitement of their own discoveries;
- interact with students from different schools who share common interests;
- view science as a dynamic force in their own lives;
- critically consider the credibility of their own research and the research of others; and
- draw upon the discoveries of other students to enhance their own research.

During the same time that I was considering the possibilities of such a conference for children, I was working with a group of teachers as part of the Elementary Science Integration Project. ESIP was developed by Dr. Wendy Saul at the University of Maryland, Baltimore County. Dr. Saul realized the importance of enhancing curricular integration, especially with science and literature, and promoting inquiry science in the elementary classroom. For several summers, ESIP teachers from a wide range of school systems had been meeting to explore such issues. It was from this group that support for a conference began to grow.

At first, invitations to other teachers and their classes to participate in a conference went unanswered. An inquiry conference that revolved around the authentic research of students seemed a bit risky for many teachers. And as with other new ideas, there were no models to follow. Eventually, however, two other teachers became convinced that their children could handle the challenge. That first year, three classes were involved: an urban sixth grade, a suburban fifth grade, and a rural fifth grade. Students in each class had been asking questions, gathering and recording data, and making discoveries. Although these students came from diverse regions, they all shared a common language of science; they were experts in their own areas of research and were eager to share with one another what they had discovered. The challenge was to create an event that met the goals envisioned for such a meeting.

The conference consisted of three sessions during which students attended an array of concurrent presentations and activities. Students shared with each other their discoveries about such topics as plants in sealed terrariums, mealworms, and the possibility of fossils on Mars. While student presentations were going on in some locations, hands-on activities provided by students were available elsewhere. Children demonstrated and assisted others in breaking open rocks to search for fossils or in studying the effects of erosion using a model mountain. Meanwhile, campus tours of the science labs were conducted by university students. Each session lasted about forty minutes. Students who presented in one session became participants during another as other students shared their own stories and discoveries. All around were kids behaving as scientists. They were sharing their data and discoveries with colleagues, asking and answering questions, suggesting possibilities, and wondering about further questions to ask and investigate. By lunch, we knew the conference had exceeded our wildest expectations.

Our keynote speaker during the afternoon was Seymour Simon, author of numerous children's science books. He had attended several of the presentations that morning and congratulated the students for the work they were doing. (Imagine being a fifth-grade student doing a presentation about the possibility of fossils on Mars with Seymour Simon in the audience). He went on to encourage the kids to continue with their research and made it clear that he considered each of them a fellow scientist.

The trip home from the conference was different from the quiet morning ride.

"The other students asked me all kinds of questions!" Amanda exclaimed. Her investigation was about the diet of mealworms. "They really wanted to know more about what I had found out"

"How did you feel?" I asked.

"Really neat," she responded. "It was like I knew a lot and they wanted to hear about what I had discovered."

Chris had been working in the hands-on area with the earthquake machine, a battery-operated device that simulated earthquakes for buildings made of blocks. "A man came over and asked about earthquakes. My eyes followed his tie up to his name tag and I saw that it was Seymour Simon. I got real nervous and couldn't believe that *he* was asking *me* questions!"

Mixed with the relief and excitement was a sense of satisfaction. Each child had done his or her best and made genuine contributions to the event, either through presenting, helping with the hands-on area, or by being in the audience and asking questions. Everyone received a certificate for attending. There were no losers. Each and every one of us came home on that bus a winner.

But that was just the beginning. The children returned with new ideas and different things to try. A science investigation is not a science project. It might not *ever* be finished. The children came back anxious to continue their research and proud of their achievements at the conference.

The year of the first Kids' Inquiry Conference was also the beginning of an informal network of teachers who collaborated on the development of KICs in a variety of venues and grade levels. Although the Elementary Science Integration Project was key to the success of the first KIC, our goal all along was to develop a model that would enable classroom teachers to independently organize, plan, and host their own inquiry conferences. Such a model is described here not as a prescriptive method to follow, but as a starting point around which future conferences can be planned, reflecting the differing characteristics of the classes involved.

Planning and Conducting a Kids' Inquiry Conference

Getting started with your own inquiry conference means thinking about it early in the year. The notion of inquiry science as a student-centered activity must be accepted by the teacher and communicated to the children from the very start. Children who have explored and investigated throughout the year are naturally ready for a spring conference. On the other hand, starting inquiry in April for a May inquiry conference is definitely not the best approach. Inquiry cannot be forced; time must be provided for it to grow slowly as the children are immersed in its processes and struggle with its challenges.

Planning a conference *is* a bit involved, but it is manageable if started early and with several other teachers equally committed.

Most conferences have been held at locations away from school. Getting out of the daily environment makes the day special. Hotels, corporate headquarters, and colleges and universities have provided settings conducive to scientific discussions. A partnership early in the year is important in arranging a conference location.

Applications to Present

By February the students are ready to complete KIC applications to present (Figure 9-1). Their goal is to convince the "KIC Committee" to accept their presentation and schedule it for the conference. The application process requires the students to tell about themselves as scientists, describe previous discoveries, and explain the nature of their current investigation. The students are also asked to include an assessment of their research thus far. Are they being successful? How do they know? This requires thinking about exactly what success might mean. (I like to remind them of the story of Thomas Edison and his search for a working "glow bulb." After trying hundreds of materials for the filament, all of which were unsuccessful, Edison finally tried carbonized cotton thread, which glowed for forty hours. Each "failure" was actually a discovery of what *did not* work. The failures eventually led to success.)

Some teachers require every student to apply. I prefer to let each student choose. Often students form small groups of two or three and apply together. Even though everyone in the class might not do presentations, the entire class will go to the conference.

When the applications are completed and turned in, the teacher sends them to the KIC Committee for review. This designation is a little misleading. The students think that they are writing for a group of people far away. Actually, each teacher is the KIC Committee; he or she evaluates the applications. Teachers sometimes exchange

KIC:
Application to Present

The Kids' Inquiry Conference Committee is eager to hear about your scientific research and discoveries. In order to plan the conference, your assistance is needed. Please complete the spaces below.

NAME ______________________________ DATE ______________

SCHOOL ____

TEACHER ____________________________ GRADE _____________

1. Describe any reports or projects that you have prepared for science class in the past two years.

   ____
   ____
   ____
   ____

2. List any science articles or booklets that you have published in recent years (such as in the class or school magazine or newspaper).

   ____
   ____
   ____

3. List two or three discoveries you have made within the past two years, either at school or elsewhere.

   ____
   ____
   ____
   ____

4. Describe the question you are researching and would like to present at KIC.

   ____
   ____
   ____
   ____

FIGURE 9-1 *Application to Present at the Kids' Inquiry Conference*

5. Briefly discuss your investigation. What are you doing to attempt to answer your question?

____

____

____

____

____

6. How is your research progressing? How would you evaluate the success of your investigation thus far?

____

____

____

____

7. What will you include in your presentation to convince your audience that your discoveries are valid?

____

____

____

____

____

Please sign below if you are willing to present your findings at the Kids' Inquiry Conference. Your application will be reviewed by the KIC Committee and you will be notified of the committee's decision.

____

SIGNATURE OF APPLICANT

____

SIGNATURE OF PARENT

____

SIGNATURE OF TEACHER

Thank You and Good Luck!

FIGURE 9-1 *(Continued)*

applications and review those of students in different classes. But it's amazing how differently a student will approach the completion of the application depending upon the audience. Knowing (or thinking) that someone other than their own teacher will decide who will be accepted to present at the conference enhances the quality of the completed applications.

The applications are evaluated based upon several criteria. I make certain each student or group of students is clearly on a path of inquiry. The conference is not a place to be presenting book reports. Are the investigations scientifically valid? Are the processes of science being applied correctly? Do the data seem reasonable, and if not, is there a credible explanation? The teacher's role is to make certain that the presentations accepted are ones from which other students will learn about science. The teacher must also make absolutely sure that no student will be placed in a position of embarrassment because of faulty test procedures or obviously inaccurate data.

Almost always, every student's application is accepted. What an exciting day it is some weeks later when the package arrives with the responses from the KIC Committee. Each student who has been accepted to present receives a letter (see Figure 9-2 for a sample). It is a great feeling to find out that someone thinks your research is worthy to be scheduled for an inquiry conference. The jumping up and down, screaming, and congratulating is soon followed by the somber realization of just what the acceptance letter means.

"I am *really* excited, Mr. Pearce," Amanda told me after reading her letter of acceptance. "But now I'm nervous. I have to stand up in front of people I don't know and talk. I've never done that before!" In the weeks to come, my job shifts a bit. It is up to me to make certain each child is prepared and comfortable and able to do their best at the conference. I reassure them that they have wonderful stories to tell of their research and their discoveries. Usually we have about eight weeks from receipt of the acceptance letters until the date of the conference.

Occasionally there are students whose applications are simply not acceptable. I never outright reject a student's proposal. Instead, the "KIC Committee" sends a letter to the student thanking him or her for applying and asking the student to respond to several questions that may help clarify the student's proposed investigation. The questions may address test procedures, questionable use of materials or the treatment of animals, or safety issues. If the student responds by a certain date, the proposal is reconsidered.

Hands-On Displays

In addition to planning presentations for the conference, students might opt to prepare a hands-on display. These displays are set up at tables behind which the host student(s) lead activities for other students browsing from one table to another. Hands-on displays have included opportunities for students to build on the earthquake machine, mix ingredients to make putty, make ice cream, test the pH

To: Jeff Miller
Date: March 21, 1997

Congratulations ! ! !

Your application to present on the topic of *salt water and grass* at the 1997 Kids' Inquiry Conference has been accepted by the KIC Committee. We are excited about your research and look forward to your presentation on May 2, 1997 at Becton Dickinson.

To plan your presentation, please consider the following guidelines.

1. Review again the *exact* question you tried to answer.
2. Read through the notes and records you kept during your research. Review *exactly* what you did to answer the question.
3. Think about the results of your investigation. Was your question answered? If not, can you explain what happened? If yes, will you be able to convince others that your results are valid?
4. Plan your presentation. It should include:
    - your original question;
    - what others have discovered about similar or related questions;
    - a description of your investigation (photographs, diagrams, and charts might be helpful);
    - the results of your investigation *(your discoveries!)* (charts, graphs, and/or tables will make your results more convincing);
    - additional questions that might be interesting to pursue later by you or others.

When you present, try to relax. Speak slowly and loud enough for everyone to hear. Repeat for your audience any question asked by individual students.

Your class and teacher are proud of you. All of us thank you for your willingness to share your discoveries with other student scientists!

PLEASE HAVE YOUR PARENTS SIGN THE PERMISSION SLIP BELOW, TEAR OFF, AND RETURN TO YOUR TEACHER.

- -

I have read about my child's invitation to present at the Kids' Inquiry Conference on May 2, 1997 at Becton Dickinson.

____

STUDENT SIGNATURE

____

PARENT SIGNATURE

A complete letter and permission slip explaining the details of the Kids' Inquiry Conference will be sent home soon.

FIGURE 9-2 *Sample Letter of Acceptance to Present at KIC*

of different materials, build and test boats, experiment with a variety of magnets, and construct hourglasses with soda bottles and salt. The best hands-on activities are those planned and operated by students. They make posters, write directions, have guest registers, and have even included Halls of Fame for visiting students who distinguish themselves. One year a group of students had a UV bead table near a window. Visitors were invited to use different materials to view the effects upon the beads. Later, the guest register was used for a random drawing. A UV bead necklace was awarded to the lucky winner.

A hands-on display is ideal for those students who may be too shy to actually present. Or, hands-on displays have been used as a follow-up to a presentation where there is more time for other students to experience the activities. Several presentations at our most recent conference ended with invitations for students to visit the hands-on display table later in the day.

Students who want to provide a hands-on display are required to apply (see Figure 9-3). Here again the KIC Committee wants to make sure that the displays are safe, manageable, and contain reasonable activities. The teacher is the ultimate judge of those applications that are submitted.

Although student-operated hands-on displays have proven to be quite popular, they need not be included in every conference. Your first attempt at an inquiry conference may best be kept simple, stressing the presentations alone.

Presentation Descriptions: Blurbs

As the conference date approaches and all the responses from the KIC Committee have been received, the students must write presentation descriptions. These are short blurbs (thirty-five words or less) that include the presentation title (as it will appear in the program), the student scientist(s), and the topic of their research. A sample blurb might be:

All Charged Up!

by Tim and Justin

> Have you ever wondered if a new battery can recharge a dead one? We have been experimenting with old batteries and ways to bring them back to life. We have even invented a cordless charger. Come to our presentation and see what we discovered.

Written blurbs are used to organize the conference schedule and to communicate to everyone attending what the presentations will be. With the blurbs in hand, it is a simple task to arrange the presentations into groups. If the conference will have three sessions with three different meeting rooms during each session, there are nine different locations for presentation scheduling. Grouping three or four presentations for each location along certain themes can be accomplished with the blurbs as a guide. We cut the blurbs into strips of paper and then mix and match on

KIC: Application for a Hands-On Display

The Kids' Inquiry Conference Committee is interested in providing a variety of hands-on activities at KIC. Students and adults learn best by doing. The KIC Committee would like to invite you to plan and operate a hands-on activity. Please complete the spaces below.

NAME ______________________________ DATE ______________

SCHOOL ____

TEACHER ____________________________ GRADE ____________

PARTNERS YOU PLAN TO WORK WITH: ___________________________

____

1. Describe any reports or projects that you have prepared for science class in the past two years. ______________________________

   ____

   ____

   ____

2. List any science articles or booklets that you have published in recent years (such as in a class or school magazine or newspaper). ______________

   ____

   ____

3. List two or three discoveries you have made within the past two years, either at school or elsewhere. ______________________________

   ____

   ____

4. Describe the hands-on activity that you would like to offer at KIC. What will be the topic? ______________________________________

   ____

   ____

   ____

FIGURE 9-3 *Application for a Hands-On Display at KIC*

5. Explain what visitors will do at your hands-on activity.

____

____

____

6. List the materials you plan to have available at your hands-on activity.

____

____

____

7. What do you hope visitors to your hands-on activity will learn?

____

____

____

Please sign below if you are willing to participate in the hands-on portion of the Kids' Inquiry Conference. Your application will be reviewed by the KIC Committee and you will be notified of the committee's decision.

____

SIGNATURE OF APPLICANT

____

SIGNATURE OF TEACHER

____

SIGNATURE OF PARENT

Thank You and Good Luck!

FIGURE 9-3 *(Continued)*

a grid that identifies session and location. This is a good task for the participating teachers to work on together, or for one of the teachers to do if distance between schools is a problem.

Sign-Up Sheets

Once the presentation schedule has been identified, a separate sign-up sheet for each room and each session must be made (see Figure 9-4 for a sample sign-up sheet). These sign-up sheets are for the students to select which presentations they will be

KIC Session Sign-Up Sheet

KIC '98
Becton Dickinson
May 18, 1998

SESSION II
RED ROOM

BACTERIA
by Jen O'Dell and Greg Miller
What is the one thing that seems to be everywhere? No, not air! It's BACTERIA! and they can wipe out anything, including humans. Come see our presentation and help us fight this threat. We're counting on you!

ACID + BASE = WHAT?
by Scott Castro
If you have you ever wondered if any other acids and bases (besides vinegar and baking soda) make a reaction then come to my presentation. I will tell what I discovered!

NATURAL ANTIBIOTICS
by Kathleen Bender
How is our forest like a rain forest? They both contain *antibiotics!* Come to my presentation to see what they are!

1. ______________________________
2. ______________________________
3. ______________________________
4. ______________________________
5. ______________________________
6. ______________________________

FIGURE 9-4 *Sample Class Sign-Up Sheet for a KIC Session*

seeing at the conference. Each class receives its own sign-up sheet for each session and room. The presentation descriptions are included for the particular time and location with a limited number of available spaces. This method ensures a mix of schools and equal numbers of students at each presentation room. For example, if three classes with a total of ninety students were attending a conference, each presentation room would accommodate thirty students. Each class could send ten students to each location (including those scheduled to present).

Name Tags

Using the completed sign-up sheets, the teacher can easily make name tags for the students. In addition to the student's name and school, the name tag includes three colored dots, one for each session. This works to minimize confusion on the day of the conference. If a student signed up for the red room for session 1, the blue room for session 2, and the green room for session 3, that student's name tag will show three dots: red, blue, and green. At a glance, it is easy to tell who is to be where.

KIC Journals

During the first conference, several students remarked on how frustrating it was that they could not attend all of the presentations. With concurrent presentations, students had to make difficult choices when they signed up. To help address this concern and to document the proceedings of the conference, the Kids' Inquiry Conference journal was begun the following year.

The KIC journal contains articles written by conference presenters describing their investigations and discoveries. Adding this writing component to the KIC process helps to integrate inquiry with other subject areas. Plus, the KIC Journals make wonderful reading materials the following fall when a new group of students begin the process of inquiry all over again. The authors of the articles serve as role models for future students.

Using KIC journals as science texts provide many opportunities for reading and writing activities in subsequent years, some of which are described in Chapter 7. In addition, the journals provide a span across the summer, linking each year's class with the next. As a tool for building community, the KIC journals are a wonderful means of maintaining and communicating the accomplishments of the student scientists.

A KIC article from a recent journal is shown in Figure 9-5. The article contains the question the student pursued and where the question originated, background information on the topic, a description of the investigation, discoveries made by the student, and some possible questions future students might investigate. These kinds of articles have provided ideas for subsequent students to attempt to replicate, or served as sources for other questions that may prove interesting. I encourage the children to be original, but also explain to them how science builds on what others have discovered. It is really exciting when a student reads a KIC article and then decides to carry the investigation one step farther. This process is authentic science, and it happens right in the classroom!

Robby's presentation included overhead transparencies that displayed the data he had gathered. He also told about problems during his research—his hamster chewed on the wires and gnawed at the magnet. He had to try several different methods before he was able to gather his data successfully.

Each fall, as we read the articles from past conferences, I like to ask the students if enough details are included for them to carry on the investigations. Sometimes the

Hamsters
by Robby Waskiewicz

My topic is hamsters. I think this is an interesting topic to research. I chose this topic because I have a hamster and I wanted to know more about him.

The questions I researched were: Is it possible to measure how far a hamster can run in one day? If so, how far does a hamster run in a day?

Hamsters are a member of the rodent family. They are related to mice, gerbils, rats, and muskrats. The word "hamster" means to hoard food. Most hamsters came from Europe and Asia. Hamsters like running in their wheels. In a book I read by Alvin and Virginia Silverstein, *Hamsters: All About Them,* it said that hamsters can learn how to walk a tightrope and go down slides. They are smart.

To do this research I took off the distance computer that measures how far you run on a treadmill and attached it to my hamster's running wheel. My dad helped me. We then calibrated the wheel by measuring the circumference of the wheel. It was 20 inches. Then we put a magnet on the inside of the wheel. Next we used Velcro to attach the distance computer. The computer reads the magnet every time it goes by. We had to measure how many times the wheel went around to make the computer click 1. It was six times. Then we took the six turns and multiplied by 20 inches (the circumference) and got 120 inches. After that we divided by twelve (one foot) and got ten feet. So each click is ten feet. I fed my hamster 10 pieces of food per day. I did this because I wanted him to have the same amount of energy each day.

I discovered that my hamster does not run the same distance each day. Do you believe that he runs a couple of miles per day? (See my chart for the distances ran each day.) I was surprised how far he ran. When changing the amount of food, there seemed to be no change in the distance.

Other students might try this research and maybe ask: How far can a hamster run in a week? I plan to continue with my research.

Bibliography

Silverstein, Alvin and Virginia, *Hamsters: All About Them,* New York: Lothrop, Lee & Shepard, 1974.

[The author also included a bar chart that displayed the data.]

FIGURE 9-5 *Article Published in the Kids' Inquiry Conference Journal*

description of the investigation is insufficient to fully understand or replicate what was done. Fifth-grade authors who have moved on to middle school are not available to fill in the missing details. Current students who wrestle with this problem are careful to include more of the necessary details when the time comes for them to write their own articles. The quality of the articles has definitely improved from year to year, in part because of what the students have learned from one another.

Organizing and funding the publication of the KIC journal can be accomplished in a variety of ways. Once the articles are written and carefully proofread, they are typed either by the students themselves or by parent volunteers. I prefer having each teacher be responsible for copying the articles of his or her own class. For example, suppose three teachers decide that they want to have one hundred journals. Each teacher would then arrange to make one hundred copies of the articles from his own class. How to cover the costs of copying can be decided upon by each teacher. Some administrators will pay for the project because of its appeal to both science and language arts. In some schools, the PTA or other parent group might fund the copying. Some teachers charge two or three dollars for the journals for their own students and use the money to make the copies. However each teacher arranges for copying, the articles are then brought together to one location, collated, and bound.

Moderators

Before the day of the conference, it is important to think about moderators who will be in the presentation rooms during each session. Moderators are adults who assist with directing children to the proper locations (based upon the colored dots on the name tags), welcome children to the room, and then introduce each presentation group to the audience. (The presenters will have prepared biographical information sheets prior to the conference to assist the moderators with introductions. See Figure 9-6.). Quality moderators are essential for a smooth-running conference. A guide for moderators is shown in Figure 9-7.

With two or three presentations during a forty-minute session, the moderator must watch the time carefully to make sure each group gets to present. Typically, a presentation will run eight to ten minutes, with additional time for questions and discussion from the audience. The moderator also helps to facilitate discussion following the presentations. Often the children in the audience will be a bit shy during the first session. The moderator should have a question or two of her own to ask the presenters to get the discussion rolling. Such generic questions as, "Where do you plan to go from here with your research?" or, "What are some questions future students might try to answer on your topic?" are helpful. Actually, the question-and-answer portion of the presentations can be the best part. The presenters are less nervous and are happy that someone in the audience wants to know more. Some wonderful discussions have occurred during this part of the presentation sessions.

The role of moderators can be filled by visiting teachers (it is best to have the classroom teachers available to float from room to room to take care of any problems that may arise), visiting administrators (who are often happy to be included and enjoy working with the kids), or parents or guests (who have been carefully selected). Conferences at colleges and universities have invited education students to serve as moderators. Serving as a moderator is a terrific experience for college students about to student teach or work in the classroom in some other capacity.

KIC:
Presenter's Biographical Data Sheet

Directions to presenters:
A. Complete the spaces on this form.
B. Give this form to the moderator in your presentation room at the conference.

Your name ______________________ Age ____ Grade ____
Teacher ______________________ School ______________________

School activities: ______________________

____

____

Hobbies: ______________________

Favorite books: ______________________

How did you get interested in scientific research? ______________________

____

____

List one or more of your accomplishments. ______________________

____

____

____

____

What other science questions have you investigated? ______________________

____

____

____

____

What are your hopes for the future? ______________________

____

____

Why did you decide to present your investigation and discoveries at the Kids' Inquiry Conference?

____

____

____

____

Thank You for Participating in the Kids' Inquiry Conference!

FIGURE 9-6 *KIC Presenter's Biographical Data Sheet*

KIC:
Moderating KIC Sessions

Thank you for agreeing to moderate a research and sharing session at the Kids' Inquiry Conference. Here are a few tips concerning the role of a moderator.

KIC was designed so that kids could share the excitement of their science research with their peers. These sharing sessions are the heart of the conference.

Most sessions will have presentations from two or three groups of student scientists. Some of the children have been in touch with one another. Others will be unfamiliar with the research of their fellow presenters.

Each session will be forty minutes long. The structure of the sessions is flexible, but should include the following:

- The moderator will introduce student presenters (don't forget to introduce yourself as well).
- Each student/group will present their research.
- There should be time allotted for questions and discussion.

Introduce each group of scientists. You may be given a brief biography of each child. If not, get their names and school in advance.

Establish a sense of order and monitor time. You are in charge of the session. It *must* be completed in forty minutes. Make sure everyone gets a turn to present.

Help with AV equipment. Students may be using overheads, slides, and VCRs. All equipment will be in the room.

Ask, "Where do we go from here?" Encourage students to think about what new questions they have after the presentation, how they could explore these questions, and/or how they could help one another's research.

Do not let children leave early, but please have everyone ready to leave at the end of the session. Time is very tight. There will only be five minutes between sessions.

There will be plenty of parents/chaperons/teachers/staff around. Feel free to use their help with discussions, questions, AV equipment, bathrooms, errands, timing, and so on.

Thank you for helping to make KIC a success!

FIGURE 9-7 *Guide for KIC Moderators*

Printed Programs

It is a good idea to have a printed program for the day. The program might include: the schedule, with presentation titles and times as well as colored-coded room locations; a map of the conference site; a description of any special meeting places; a listing of the classes and schools attending; and a list acknowledging volunteers who helped make the conference possible. Also included might be space for students to write names and e-mail addresses of others they meet.

The Day of the Conference

After many months of anticipation, the big day finally arrives. It is normal for teachers to feel a bit apprehensive. Can their students really do this? Are they adequately prepared? Will they tell their stories in front of strangers as well as they did back in the classroom during the practice runs? After so many years of taking my students to inquiry conferences, I still feel a bit anxious before they start. Every year, however, I have found that the children truly do their best and often surprise me (and their parents) by the way they handle themselves in front of an audience.

The buses arrive and the children unload their boxes and bags of materials. The presenting students are escorted to their presentation rooms to organize their materials. Then everyone assembles in a large room or auditorium for welcoming comments. Participants are then dismissed to their respective session 1 stations, and the children are off on their adventures.

Following the first two sessions, lunch, and then session 3, the children reassemble in the large meeting room for the keynote address. Speakers have included visiting authors, local scientists, administrators, teachers, and college professors. Whoever the speaker may be, their message should include an authentic science story or two, congratulations to the students for having the courage to attend the conference and present the results of their investigations, and encouragement for the children to continue on the path of inquiry and discovery. By the end of the day the children all think of themselves as serious scientists. Hearing this confirmation from an adult speaker sends them on their way knowing that they are part of a wider scientific community.

As the day ends and the children board the buses to return home, the feeling of accomplishment is palpable. For most, this day has been their first time standing and speaking before a group away from school. They have told their stories, shared their data, displayed their charts and diagrams, and have contributed to the scientific body of knowledge. It is on the bus during the ride home that I distribute the KIC journals. The children relive the day by looking first to find their own article, and then looking through the journal for articles by students from the other classes they met. That evening there are many stories to tell about their day at the Kid's Inquiry Conference.

Conference Follow-Up

The investigations do not end when the conference is over. Many of the students continue to gather data. Others decide to change their question in some way as a result of the discussions at the conference. Those questions that future students might try to answer are sometimes too compelling to leave for others.

Robby's investigation (as described in his article shown in Figure 9-5) concerned the distance his hamster ran on a wheel each day. His audience at KIC wanted to know about the effect that different foods or sugar additives might have on the hamster's performance. Some were wondering about different types of hamsters and which might tend to run more than others. After the conference, Robby was interested in trying other things that he hadn't thought of before and planned to continue his research during the summer.

Most years, our conference takes place near the end of the school year. Conducting continuing research at home makes for an interesting summer for the students. The year that Kayce spent researching antibiotics that might be found in plants in our outdoor classroom continued as summer approached and the forest became lush with vegetation. Kayce did not want to stop her inquiry merely because the school year was over. Fortunately, we had many materials for Kayce and others to take home for the summer. Their research continued (although I'm sure their parents wondered about the strange odors coming from their rooms).

Post-KIC surveys (an example of one is shown in Figure 9-8) have proved helpful to assess the value of the day and to see how the students grew as a result of the experiences at the conference. About the conference and inquiry research, students have made some rather profound remarks:

> "Sometimes friendship gets in the way of research."
> "Your age doesn't limit your scientific ability."
> "Listening is really important."
> "When I was doing my research I learned that a scientist does a lot of work, but it can be a lot of fun, too."

Children's discovery of the difficulties and rewards of science can be just as important as their revelations about the world around them. Perhaps Jessica's journal entry, written the day after she went to KIC, summed it up best.

> I'm very glad we got to have KIC because it really was a lot of fun! It was cool to see other kids' presentations/experiments. It was also a lot of fun to do my own presentation. My favorite part, however, was the hands-on time. It was fun to see people try my hands-on activity.
>
> I think KIC would be very important to have again next year. I say that because KIC gets kids' minds and imaginations going. We start to do more experimenting on our own. I think this is a good thing (and it keeps us out of trouble)!

Helping kids to become lifelong learners means guiding them toward independence. Once they become self-motivated again (just like they were as toddlers) a

Thinking About KIC

Name ________________________________

Please think about your experiences preparing for and attending the Kids' Inquiry Conference. Answer the questions below.

(Circle any that apply): I was a *presenter* *hands-on participant*

1. What did you do to prepare for KIC?
2. What were your thoughts while you attended KIC?
3. Did preparing for and attending KIC make you a better scientist? How do you know?
4. How has your thinking about being a scientist changed because of KIC?
5. What do you now think about science that you didn't think before?
6. Should KIC be held again next year? Why?

On the back: Please write the most important thing you learned from participating in KIC.

FIGURE 9-8 *Thinking about KIC: A Post-KIC Survey*

part of our task is largely done. It can be exciting and satisfying to see the spark of inquiry rekindled.

Frequently Asked Questions About KIC

Q. Should I require every student in my class to prepare a presentation?
A. The value of KIC lies in kids' ownership of the experience. Ideally, each child will view the conference as a chance to share. Students who are excited about doing something for KIC become self-motivated because they are invested in their projects. However, busy classrooms that offer numerous choices to children may

provide alternative endeavors. I prefer to let each child decide if they want to present, provide a hands-on display, or merely attend the conference as a participant. There are also those students who may decide to do as little as possible when given the choice. More structured assignments are given to them during class KIC preparation time.

Q. What are some things we can do in our classrooms to help conference participants get to know one another before the conference?
A. Some class members become pen pals or e-mail buddies prior to the conference. This communication is helpful, since it enables the students to share ideas as they prepare their presentations. It is also fun to meet someone at the conference after months of corresponding.

Another possibility is a video exchange. One videotape, passed along and added to by each class, can include the students introducing themselves and briefly describing their investigations. After going to each class, the tape is then sent back to the original teacher, so that all of the students can eventually see one another. The children enjoy seeing others who will be at the conference, especially if they have been pen pals during the previous months.

Another idea is for the classes to exchange the topics being researched. We have had success with writing each topic on newsprint and posting them on the walls around the room. Students then visit each topic and add questions of their own. The papers are sent back to the classes so that the students preparing their presentations have an idea of what their audience might like to know about the topic. This can be useful in deciding what to include in the presentation.

It is important to remember that as we build a community inside our classrooms, KIC offers an opportunity to include our students in a wider community beyond.

Q. When during the conference are the hands-on displays scheduled?
A. There are several options here. One is to schedule the displays during an extended lunch period (about one hour). When the students are finished eating, the displays are available for browsing. This works well because it gives the students something interesting and fun to do when they are finished eating and waiting for the next session to begin.

Hands-on displays can also be scheduled during one or more of the sessions. The problem is that not everyone is available to either run or visit the displays.

Q. Other than presentations, what other activities can be scheduled during each session?
A. This depends upon the location of the conference. At colleges and universities, tours of laboratories have been conducted by education students and professors. At some schools, professors have planned special lab activities for one or more of the sessions. Corporate locations have also included tours of the facility. By scheduling other activities during each session for which students may sign up, group sizes in the presentation rooms can be more manageable.

Q. Should I assign questions for my students to investigate for KIC?
A. No! The students' own investigations are the most interesting. If you notice a child who has a particularly engaging question, you might encourage him or her to continue with the idea with KIC in mind. Inquiry is for all students, and therefore there will be different levels of complexity in KIC questions. Some questions may seem simple to us as adults, but even the simplest question allows for scientific discovery. In 1996, Luke was interested in air power. He wanted to examine the effect of air movement on different sail-like materials. He devised an experiment in which he used a handheld fan to propel different kinds of sails across a string. He tried paper, cardboard, and other items and measured how far each traveled. My first response might have been to simply tell him that one material was better than the other, but that was not the point. Luke learned *how* to do science by using ordinary objects to pursue his own question.

Q. How does an investigation presented at KIC differ from other scientific experiments in my class?
A. Investigations presented at KIC are those originated and selected by the students themselves. Many experiments in class are part of the science curriculum, which is teacher centered. These experiments have known results and are performed by the entire class together. Those experiences are valuable to help the students understand the processes of science and to provide the prior knowledge on which to later build. But the investigations presented at an inquiry conference represent actual research in which the answers are *not* known at the outset. Even if a student chooses to replicate a curriculum experiment or an investigation of another student as described in a KIC journal, the results are not predetermined. The difference between a class experiment and a KIC investigation is the ownership. Once ownership is assumed by the student, the investigation takes on an entirely different identity.

Q. How can I plan my own inquiry conference without an organization like ESIP to help me get started?
A. The Elementary Science Integration Project at UMBC provided vital support in getting the Kids' Inquiry Conference started. ESIP personnel paved the way for the first conferences by arranging meeting locations, handling clerical demands, and being available to help make each conference day a success. The goal from the start, however, was to provide a model that could evolve and survive on its own. Each year, ESIP played less of a role as the conference spread to other colleges, universities, and corporate locations. In recent years, teachers beyond the ESIP network have successfully conducted their own inquiry conferences, using what was learned earlier and implementing their own innovations. The two most important aspects of planning your own conference are a group of three or four teachers committed to providing an inquiry conference, and a deep-seated confidence in your students' ability to share what they have discovered. With these in place, your inquiry conference will surely be a success!

SECTION III
Assessment

Ten

Tools for Assessment

Assessment is a continual process. It begins the moment we meet our students at the start of the year and continues to the last day. I assess throughout each day, constantly looking for evidence of progress.

As a teacher, I ask myself three basic questions: What do my students know? What are my students able to do? What else do I want them to know and be able to do? To assess I must carefully observe and have a clear idea of where I think my students should be as the year progresses.

Assessment of student progress in inquiry science requires several different approaches. In one respect, inquiry science is easily assessed. The children are engaged in many activities developing tangible products. Observing their behaviors and examining their products helps to determine progress attained.

On the other hand, since an inquiry approach leads toward student autonomy and independence, children working in diverse areas may be a challenge to assess in traditional ways. Nontraditional assessment and student self-assessment take on greater significance.

The Behaviors of Inquiry

From the start of the year I know exactly what I hope to see in each student. Even though inquiry leads to individual autonomy, the teacher *must* be aware of where the students are going and how they are getting there. Clearly defined, observable behaviors that indicate an understanding of the processes of inquiry science must be in place in order to assess.

Throughout the year I ask myself questions about each student. The questions represent some of the most important indicators that I use to evaluate the progress of my students and the success of my own role in the classroom.

Is the Student Asking Testable Questions?

I recall a conversation a fellow teacher and I had in which we tried to find ways to assess questions. If we truly believe that a student's own question is the best question, how can we evaluate an authentic question from a wondering student? We decided that if a question is one that leads a student to an investigation that yields real data, which is observable and which can be utilized to explain something that was previously unknown, then the question was a quality one. (Refer to Figure 2-3 for ways to help students phrase questions that may lead to more meaningful investigations.)

Does the Student Design Fair Tests to Answer Questions?

Children are always trying things to see what will happen. Younger children will often have difficulty controlling variables. A fair test is one in which only one variable at a time is changed in order to determine correlation and, in some cases, causation.

Does the Student Gather Data in an Organized and Logical Manner?

Without data, science cannot exist. Even though children have been gathering data since birth, there is a definite need for guidance and then practice in the ways to effectively collect and organize information. I look at the data my students are gathering and ask for explanations as to how it might be important and how it might relate to other information.

Does the Student Read for Additional Information Related to an Investigation?

Being aware of the books students are reading helps me to assess their interest in and commitment to an investigation. Those students who seek books from the library or ask for my assistance with books in the class library communicate to me a desire to find out what others have discovered and how that information might relate to their own investigation.

Is the Student Able to Exhibit an Understanding of Variables?

An independent variable is a change made during the course of an experiment. Dependent variables are the result or effect of such a change. For example, when experimenting with the bounce of a ball on different surfaces, each surface would be an independent variable and the height of bounce would be the dependent variable. (The ball, remaining unchanged for each trial, would be a controlled variable.) Some younger students may have difficulty distinguishing types of variables. Calling an independent variable a "manipulated variable" may help. Although observing a correlation is far easier than assigning causation, a student who explains that doing one thing leads to another (cause and effect) understands that effects can be influenced by certain actions.

Is an Understanding and Use of a Control Exhibited by the Student?

As students manipulate variables in an investigation, the proper use of a control indicates a maturing sense of a fair test. A control is utilized to rule out one or more possible causes in a set of results. (See *Inquiry In Action: The Search for a Natural Antibiotic* in Chapter 8 for an example of a student use of a control variable in an investigation.)

Is the Student Able to Translate Observations into Usable Data?

Children observe the world around them from birth. A camera also observes. The difference is that a child can take what is observed and record the observation as data to be used at a later time. A child observing an insect and noting the colors of the flowers that insect visits is using her observations to record data. When a student tells me about an observation, I will often ask, "How do you think that observation might be important?" or "What does this observation tell you so far about what you are observing?"

When Joe made the observation that shaking the UV beads made them change colors, he was gathering data that was useful to explain what was happening to the beads. The fact that his conclusion was incorrect did not diminish the correlation he noticed. Joe, along with countless other scientists through the ages, simply did not have enough observations and data to accurately explain causation.

Does the Student Discuss Ongoing Investigations with Others?

The sharing of observations and investigations begins during and following the first inquiry period early in the year, when each student has an opportunity to tell the class about his or her activity. During subsequent inquiry periods, I look to see that students who are working together are discussing what they are doing. Listening to their dialogues tells me much about their thinking and how well they are working together. Later in the year, students may wish to tell their stories to reading buddies or apply to present at KIC. I look to see that our own scientific community allows for and enhances the flow of information between students.

Does the Student Show Perseverance, Especially During Investigations with Unexpected Results?

Maturing scientists grow more committed to the investigations in which they are involved. This means a willingness to invest time and energy, even when the results are disappointing or unexpected. Students who have difficulty staying with an activity or investigation may require more options or guidance. Of course, even the best of scientists will abandon some investigations.

In a sense, inquiry science is like reading books. We encourage our students to give each book they select a chance, reading far enough to see if the book might eventually be interesting. This attitude is modeled when I read aloud and tell how some of my favorite books were slow starters. Had I abandoned them early, I would have forever missed the unread pages.

Scientific investigations are similar. The tedium and monotony that may accompany some investigations can often lead to discoveries later. Tales of earlier students' frustration and disappointment (the fly group, those involved with antibiotic research, and others) which eventually led to wonderful results may encourage current students to press on and not give up.

I look especially for students who may be having difficulties with their experimental design yet follow through with their investigation. A great deal of problem solving occurs in overcoming the challenges of gathering data. Many students tell me (either orally or through journal entries) about time spent away from school working on their investigations. I expect for my students to be increasingly persistent as the year progresses.

Does the Student Record Data for Future Use?

Students who are gathering data are encouraged to record it. I look at discovery log sheets (in the discovery box folders), contract journals, and dialogue journals. Since we might not know which data may be needed, it is important to record as much as possible. I look to see that the students are writing lists, recording numbers, making charts, diagrams, sketches, tables—all in a location that can be accessed later.

Does the Student Ask New Questions Based Upon Newly Acquired Data?

The flow of questions is the fuel that runs inquiry. I look to see if students ask questions only in response to an assignment or if they are continually asking questions on their own. The question board is a great source of assessment data, as is daily conversation.

Is the Student Engaged in Self-Directed Investigations?

Developing natural scientists who are lifelong learners means moving toward autonomy and independence. Students who engage themselves in their own investigations based on genuine interests and who have the knowledge and skills necessary to conduct such investigations indicate to me their progress toward internalizing the values of science.

Does the Student Make Entries in the Book of Discoveries?

As discoveries are made, no matter how seemingly small, the children are encouraged to add to the Book of Discoveries. As an assessment tool, the book is a documented record of the kinds of questions and discoveries being made, how the student made the discovery, and a sketch that adds details. Our goal, of course, is to have the children make entries about experiments that are detailed enough for someone else to be able to replicate the experiment. The entries indicate to me how well the student is able to communicate the scientific concepts gained through inquiry.

Does the Student Show Interest in Replicating the Investigations or Experiments of Others?

Sometimes students become so intrigued by an earlier student's investigation that they decide to replicate what the earlier student had done. An interest in doing so indicates to me that the students read and understood what was accomplished earlier. Furthermore, as a student replicates another's work, he or she will often develop original questions in the process. These behaviors contribute to the sense of community that grows during the year.

Does the Student Make Connections Between Different Investigations?

The spiral of inquiry builds upon itself. Scientific progress has not been made in isolation. Connections are vital. So it is with investigations in the classroom. I am always looking to see if the students are using the results of one investigation as they conduct another or if students are noting similar or different results when comparing data. I look for students mentioning KIC articles or the names of other students when describing their own investigations or what they have read in books. As we examine the sequence of an investigation, I look to see if students used the results of other investigations to influence the direction of their own. Science as a social activity requires connections between scientists and their investigations. This is important to look for in our own students.

Checklists

The indicating behaviors are important clues that help me determine the progress of my students. Ongoing assessment is enhanced by use of the Inquiry Science Indicators checklist, as shown in Figure 10-1; this incorporates the questions from above with several other indicators. Checklists are helpful in recording what the children are doing, especially over time. How is their observed behavior changing? In what areas are they progressing?

Using the list helps me to keep track of those behaviors crucial to inquiry success. The checklist may span a period of several weeks or months. It takes that long to observe all of the items. For most children, one checklist is used for the entire year, updating as the year progresses.

These data are useful for parent conferences. They are also helpful as I assess myself and what I am encouraging the students to do in the classroom. For example, if I notice that students are not recording their data, or not comparing the data that *is* recorded with others, I might want to do a whole-class minilesson on the importance of writing down data and procedures. (Telling the class the story of Kayce and her frustration with lack of records might help.)

Inquiry Science Indicators Checklist

For ______________________________

This student:	OFTEN	SOMETIMES	SELDOM	NEVER
1. Asks testable questions	____	____	____	____
2. Designs fair tests to answer questions	____	____	____	____
3. Gathers data in an organized and logical manner	____	____	____	____
4. Identifies and seeks additional materials	____	____	____	____
5. Reads for additional information related to an investigation	____	____	____	____
6. Exhibits an understanding of variables in an experiment	____	____	____	____
7. Exhibits understanding and use of a control	____	____	____	____
8. Translates observations into usable data	____	____	____	____
9. Discusses ongoing investigations with others	____	____	____	____
10. Exhibits perseverance, especially on investigations with unexpected results	____	____	____	____
11. Compares data with others doing similar investigations	____	____	____	____
12. Records data for future use	____	____	____	____
13. Asks new questions based upon new data	____	____	____	____
14. Creates or modifies models	____	____	____	____
15. Engages in self-directed investigations	____	____	____	____
16. Makes entries in the *Book of Discoveries*	____	____	____	____
17. Makes connections between different investigations	____	____	____	____
18. Expresses interest in replicating the investigations of others	____	____	____	____

FIGURE 10-1 *Inquiry Science Indicators Checklist*

Checklists could be coded to provide quantitative data. If the "oftens" were given a value of three points, the "sometimes" two, the "seldoms" one, and the "nevers" zero, a total score could be obtained. (For some statements the response values might have to be reversed). For these data to be valid, however, each statement would have to be of equal importance. Although interesting to try and examine, I prefer *not* counting points. Quantifiable records are at times too restricting and there is always the risk that the results of a formula will blindly lead to inaccuracy in our assessment. If formulas *are* to be used, they should be tempered with the results of other tools.

According to the *The National Science Education Standards* of the National Research Council, assessment should place less emphasis on measuring discrete scientific knowledge, determining what students *do not* know, and end-of-term evaluations by the classroom teacher. The standards encourage placing increased emphasis on assessing that which is most highly valued, evaluating scientific understanding and reasoning, and determining what the students *do* understand. A greater emphasis should be placed on involving the students in ongoing assessments of their own work and that of others. With these recommendations in mind, it is important to think about our own assessment tools, which go beyond the traditional tests at the end of the unit.

A word about assessment and grading. The toughest part of a teacher's job is translating observed behaviors and student products and progress into grades. I don't believe there exists an absolutely objective way to do this. Further, the process is clouded by our relationships with the children. How can we be coach and mentor and also judge? Gaining the trust and confidence of our students can be undermined by the realization that we possess the enormous power of grading. Balancing our roles in the classroom can be a challenge. Many curricula include assessments that must be utilized by the teacher. These are good sources for formal grades. Grading and assessment in inquiry are not necessarily synonymous. Effort and progress *must* be factors as we evaluate each of our children and maintain their trust in us as colleagues.

Kid Watching

Early in the year, it is important to determine what our students can do. Some will come to us experienced in an inquiry approach toward science learning. For others, it will be a new adventure to close the text and venture out into the unknown, uncharted waters of inquiry. Those first several inquiry periods tell us a lot about how our students think about and do science.

"Kid watching" provides important data, especially in an environment of divergent activity. I really enjoy and find useful my visits to each group of students during inquiry periods, when I can stop and listen to their conversations and get a sense of their thought processes. Occasionally I will carry a clipboard, but most often I gather notes mentally to avoid the appearance of evaluating. Casual discussions are relaxed and natural.

Other Assessments

Teachers assess for different reasons. Sometimes, assessments guide planning, helping the classroom teacher to decide what the children may particularly need. These assessments often occur early in the year or at the start of a new unit or topic of study. Other assessments help the classroom teacher decide how well progress is being made. These assessments generally occur later in a unit and are more summative. Teachers examine their original objectives and use them to assess their own success as well as the progress of the students. Pretests and post-tests are the traditional form of these types of assessment.

Often, the formal curriculum will contain conventional tests of content. These tools can provide useful information, as long as their results are only minimally used for grading purposes. Traditional tests can be more of a learning experience when administered as open-folder tests. Rather than a test of memory, open-folder tests challenge students to find pertinent information. And knowing that their folders will be available during an upcoming test, the students have authentic reasons to keep their folders up to date.

With inquiry-based instruction, the classroom teacher is fortunate to have so much more to observe and evaluate. Let's examine some of the forms shown earlier and how they might be used to assess student progress.

Student Survey

Surveys contain questions for which there are no right or wrong answers. As a result, they are excellent tools to determine student attitudes. I use this survey on several occasions to determine changes that occur throughout the course of the year. (See Figure 2-1.)

Science and Questions

This form is a type of survey. As such, there are no right or wrong answers. However, the responses can give insight into what areas of previous science study the student remembers and what areas might be particularly interesting for future investigations. This form provides valuable information that helps me to get to know each student. (See Figure 2-4.)

Kids' Inquiry Conference Journal Article Evaluation

This activity combines the reading of a KIC article with the reader's response. The reader assesses the communication skills of the author. I assess the skills of the reader to extract the question from the article and to retell the steps taken by the author to complete the investigation. (See Figure 2-5.)

Science Discovery Log

On this form, students record their experiences with discovery boxes, from which I assess a student's ability to: ask a testable question, explain what was done to answer the question, make a sketch (with labels) of some aspect of the experiment, and explain what was discovered as a result of the question and experiment. Measurements and quantities must be included. Once the students understand how important it is to properly document, I expect each discovery log page to be fully completed after each inquiry period. (See Figure 4-3.)

Scientific Discovery Form for the Book of Student Discoveries

Since making an entry into the Book of Student Discoveries is optional, doing so expresses a student's interest toward contributing to our scientific community. I look to see that the student's entry is clear and with enough details for someone else to replicate the experiment. (See Figure 4-5.)

Life at the Outdoor Classroom

This form provides another opportunity to assess the student's ability to include details in a description and information in a sketch. Since this tool is used to document life in an outdoor setting and then have the documentation included in the Outdoor Classroom Life book, the student is also writing to persuade. In order to have the page included in the book, the committee must be convinced that the student actually saw what has been described and verified it using a field guide. Students who are on the committee assess the student submitting the entry, not for a grade, but for credibility, which is the most authentic of scientific assessments. (See Figure 5-2.)

Inquiry Grant Proposal Application

Convincing others to provide funding for scientific research has become a necessary skill in modern science. To be awarded grant money, one must persuade others that the research is worthy of consideration and shows potential for some degree of success. By completing a grant proposal, students must self-assess and contemplate all aspects of their investigation, from budgetary concerns to scheduling to an evaluation of their own efforts. Assessment occurs as these forms are reviewed by the grant committee. Results of the assessment are not shown in grades but rather in dollars. If the committee is convinced, the student receives the grant. Inquiry science assessment in real-world terms doesn't get much more authentic than this. (See Figure 7-4.)

Inquiry Investigation Plan

This form is used for the student to document his own plan for upcoming inquiry periods and for the teacher to be aware of what the student plans to do. As an assessment tool, the teacher uses the information to monitor student progress as outlined. (See Figure 8-2.)

Contracts

Before I sign a contract with a student, I first assess his or her probability of success. If a student is already too busy with other activities or if he or she has had difficulty completing assignments in the past, I will advise that they wait until they have less going on. Once the contract *is* signed, I assess its progress and completion by the provisions listed on the contract itself. Those provisions should be clear so that the student can self-assess throughout the life of the contract. (See Figures 8-1 and 8-4.)

Application to Present at the Kids' Inquiry Conference

The KIC Committee (typically the classroom teacher) is responsible for reviewing the applications (to present or provide a hands-on display). (See Figure 9-1.)

Some assessments do lead to formal grades. Often, if a student does well on a particular activity or assignment I will record the results for a grade, since evidence of progress is present. If a student does *not* do well I might record the results, but not use them for grading purposes. (I am *always* looking for evidence of progress, not evidence of failure, and I communicate this to my students.) Noting a student's weak areas is important. Penalizing the student for something he cannot do now (but may be able to do next week or next month) is not supportive.

Providing ongoing feedback for the students is essential so that each can be aware of his or her own progress. Paul, for example, wanted to do everything. He found the choices in our classroom intriguing and was fascinated with a number of possibilities. He joined the Nature Table Committee and also got two contracts, one to conduct an investigation with magnets and another to write articles for the Outdoor Classroom Newsletter. Paul was having difficulty adequately recording the data he was gathering in his magnet study. When I read his contract journal, I noticed some graphs that were confusing and some gaps in his description of the investigation. A poor grade would have inhibited Paul from selecting another inquiry contract in the future. He and I discussed the strengths of what he had accomplished, and I explained how a table might be constructed to properly display his data. (Even though he had learned about tables in math, this was his first real opportunity to create one for authentic data.) A week later Paul turned in a contract journal that was greatly improved and contained real evidence of progress. Since I was confident Paul had done his best on his first attempt, the new contract journal was the one I assessed for a grade.

Performance Assessment

Assessment includes an evaluation of what the students know as well as what they are able to do. Increasingly, standardized tests are including a performance component. In Maryland, the Maryland State Performance Assessment Program (MSPAP) consists of a series of tests for students in third, fifth, and eighth grades. In the science portion of the tests, students typically read a selection about a topic and then work in small groups to gather data through an investigation or an experiment related to that topic. Then, based upon what the students read, their experiences, and the data collected, the students respond in writing to a series of prompts. Student results indicate a variety of science outcomes.

Since performance assessment tasks contain reading and writing components along with data-gathering activities, they create situations that simulate authentic scientific investigations. As a formal assessment tool, performance assessment has been able to measure far more than the traditional standardized tests.

Inquiry-based instruction in science is performance oriented. As such, performance assessments can occur in authentic ways. The reading component is present with trade books, KIC journals, and student-written log sheets. The data-gathering element is in place as students plan and conduct their investigations and experiments. And of course, writing permeates the inquiry process as students record data, procedures, and other information in journals, articles, and log sheets. Herein lies authentic performance assessment, not as part of a task to simulate the scientific process, but as real science in a community of scientists. The teacher is able to determine what the student knows *and* what the student can do.

Students Assessing One Another

The most critical assessment in science is credibility. A scientist must convince her peers that the data being reported are accurate. Being skeptical and critical are important traits that scientists share.

In the classroom, young scientists are likewise faced with the task of convincing one another. As discussions unfold or formal articles and presentations are prepared, a student's primary concern is often, "Will others believe me?" One way of addressing this question is to examine what may or may not be convincing to the student herself. Assessing the credibility of past KIC articles helps the students to see that data and supporting information from other sources are important in making a persuasive argument. Simply asking, after an article is read, "Do you believe it?" and then "Why or why not?" requires the student to examine what he or she believes and the factors that lead toward his or her acceptance of an idea.

When we go to the Kids' Inquiry Conference, the children are not worried about a grade. They want to stand before the group and have the audience believe what is presented. An essential part of preparing for KIC are the practice presentations in our classroom. Students complete evaluation forms (Figure 10-2) for one another. This information is essential as the children assess themselves as presenting scientists.

KIC Presentation Evaluation

Presenter(s) ____________________________
Topic ____________________________ Date __________

Please circle.

Rate how well the presenter(s):	effective			not effective
1. described their question(s)	3	2	1	0
2. shared background information on the topic	3	2	1	0
3. explained the investigation	3	2	1	0
4. told about their discoveries	3	2	1	0
5. suggested additional questions for future students	3	2	1	0
6. used overhead transparencies, charts, graphs, or diagrams to show data	3	2	1	0

7. Did you find this presentation convincing? Why or why not?

8. What do you think was especially *good* about this presentation?

9. How could this presentation be improved?

10. What additional question(s) do you still have about this topic?

FIGURE 10-2 *KIC Presentation Evaluation for Classroom Practice*

Students can also assess one another through classroom committees. Children who help evaluate the merits of grant requests (see Chapter 7) must think carefully about the information on the application. What kind of knowledge will be gained by approving a request and funding an investigation? How will the data be gathered? What is the likelihood of success? As the children go through the grant approval process from the other side, they are actually preparing themselves for the completion of their own grant requests.

Likewise, children serving on the committee that reviews additions to the Outdoor Classroom Life book (Figure 5-2) must assess the validity of the information.

These students learn so much about the factors that influence perception: an abundance of accurate data properly displayed, reference to the findings of others, sketches and diagrams, and careful spelling. After reviewing what is submitted by others, the students come to understand what they themselves must do to be persuasive. These assessments extend across the curriculum.

As our children assess one another, they come to better assess themselves. Guiding them to do so is our vital role if we truly want each of our students to become lifelong learners.

Afterthought

Inquiry seems so simple and natural, perhaps because as children we experienced the thrill of discovery through our own play. For many of us, that excitement continues into adulthood in a childlike way. Looking under rocks in a stream or watching a thunderstorm build on a summer afternoon still captures my imagination, and I want to know and do more.

When my own children were young, I had the perfect excuse for doing some things that I might not otherwise have done. Putting a sail on a wagon to see if it will cruise across the driveway isn't quite the same without children to help, although I still find the prospect enticing. Working with students in school and allowing them the time to pursue such endeavors makes my job so rewarding. The possibility of grandchildren who will someday open for me new areas of play and exploration is eagerly anticipated.

As teachers, it is important that we maintain a sense of wonder about the world around us. Not only should we take time to observe and question, but we should share our enthusiasm for discovery. When we model joy in exploring, we validate children's natural tendency to question and seek answers. In some ways, an adult scientist is someone who never outgrew the question-asking stage.

Inquiry is as old as our species yet as modern as the new millennium. As a means of science instruction, it will increasingly be scrutinized and discussed. The term *inquiry* may become politically charged as the autonomy and child-centered nature of an inquiry approach evokes cries to "return to the basics" by those who misunderstand its use and value. Since questions are the heart of science, facilitating questioning, experimental design, and documentation should be the core of every science curriculum. Inquiry is as basic as it gets!

At the end of each school year I look about the classroom and see the remnants of another year of student explorations and discoveries. Bits and pieces of their tests and questions remain—a fan to power a sailboat, calcium carbonate crystals in a petri dish. I can't help but play with the sand for a few moments or examine a forgotten plant that was left behind. Already I begin to think about next year. What will we do? What discoveries will be made? How will it all fit together?

I have come to expect the unexpected and to look forward to new adventures. Although so much is different from one year to another, there are those constants upon which I know I can rely: my new students will want to learn (in one manner or another), they will come with a wide range of curiosities and interests, and each and every one of them will come as a scientist. They will think in ways that scientists think, say things that scientists say, and do things that scientists do. And because of this, the adventure of inquiry will go on.

References & Resources

Professional Resources

AMERICAN ASSOCIATION FOR THE ADVANCEMENT OF SCIENCE. 1994. *Science for All Americans: Project 2061*. New York: Oxford University Press.
A guide toward science literacy for all students by enhancing science, mathematics, and technology education.

AMERICAN CHEMICAL SOCIETY. *Wonder Science*. Washington, D.C.: American Chemical Society.
Monthly science magazine with unique hands-on activities which are easily extended into inquiry.

BARRETT, KATHERINE. 1986. *Animals in Action*. Berkeley, CA: Lawrence Hall of Science.
This book launches children into authentic data gathering with classroom animals. A wide range of possibilities are discussed with sample data sheets.

———. 1987. *Mapping Animal Movements*. Berkeley, CA: Lawrence Hall of Science.
An excellent companion to *Animals in Action*, this book will help students observe and record the behaviors of several common classroom animals with practical tips and useful data sheets.

BOSAK, SUSAN V. 1985. *Science Is*. Ontario: Scholastic Canada.
Resource for numerous investigative ideas on a wide range of topics.

BOTTLE BIOLOGY PROJECT. 1993. *Bottle Biology*. University of Wisconsin. Dubuque, IA: Kendall/Hunt.
A guide to using empty two-liter bottles for a wide variety of plant and animal activities.

BOURNE, BARBARA, AND WENDY SAUL. 1994. *Exploring Space*. New York: Morrow.
Astronomy can be hands on and inquiry oriented. Here is a guide to outer-space-related activities, with references to Seymour Simon's space books as resources for prior knowledge.

BRANDT, RONALD S., ed. 1992. *Performance Assessment*. Association for Supervision and Curriculum Development.
Practical models for authentic assessment are accompanied by background information on the evaluation process.

CORNELL, JOSEPH. 1979. *Sharing Nature with Children*. Nevada City, CA: Dawn.
The classic guide to taking children outdoors.

HOGAN, KATHLEEN. 1994. *Eco-Inquiry*. Dubuque, IA: Kendall/Hunt.
Contains great ideas and resources for getting kids started with inquiry outdoors.

KAUFMANN, JEFFEREY S. 1989. *River Cutters*. Berkeley, CA: Lawrence Hall of Science.
A terrific guide to lead students toward classroom investigations in erosion. Practical and clever ideas.

MARZANO, ROBERT J., et al. 1988. *Dimensions of Thinking: A Framework for Curriculum and Instruction*. Association for Supervision and Curriculum Development.
Supports the premise that thinking as a process can best be taught when teachers provide opportunities for students to engage in different modes of thinking. A book that encourages teachers to reflect on their thinking as they plan.

NATIONAL RESEARCH COUNCIL. 1996. *The National Science Education Standards*. Washington, D.C.: National Academy Press.
A detailed description of what our students need to know and be able to do to be science literate. Includes standards for science teaching, professional development, assessment, content, program, and science education systems.

NATIONAL SCIENCE TEACHERS ASSOCIATION. 1986 (January). *The Science Teacher*.
Code of ethics for working with animals in the classroom.

NOYCE, RUTH M., AND JAMES F. CHRISTIE. 1989. *Integrating Reading and Writing Instruction*. Boston: Allyn and Bacon.
Provides ideas for scaffolding science writing and encouraging data collection.

OSTLUND, KAREN L. 1992. *Science Process Skills: Assessing Hands-On Student Performance*. New York: Addison-Wesley.
Process skills, common materials, and useful data sheets are combined to provide multilevel performance assessment activities.

PARRELLA, DEBORAH. 1995. *Project Seasons*. Shelburne, VT: Shelburne Farms.
Countless ideas for outdoor explorations based upon the seasons of the year.

PARSONS, LES. 1994. *Expanding Response Journals*. Portsmouth, NH: Heinemann.
This book is a useful reference for teachers using writing to learn strategies.

SAUL, WENDY, et al. 1993. *Science Workshop: A Whole Language Approach*. Portsmouth, NH: Heinemann.
Science Workshop contains the stories of several elementary classroom teachers who implemented an inquiry approach to science instruction in their classrooms. See the chapter entitled "What If . . . ?" by Charles Pearce for additional information about the use of discovery boxes.

SAUL, WENDY, AND JEANNE REARDON, eds. 1996. *Beyond the Science Kit*. Portsmouth, NH: Heinemann.
See chapter entitled "Eco-Mysteries" by Twig George. See chapter by Barbara Bourne entitled "The Kids' Inquiry Conference: Not Just Another Science Fair."

SEARFOSS, GLEN. 1995. *Skulls and Bones*. Mechanicsburg, PA: Stackpole.
Technical information about identifying, collecting, and safely preserving bones.

SLATTERY, BRITT ECKHARDT. 1991. *WOW! The Wonders of Wetlands*. St. Michaels, MD: Environmental Concern.
This handbook is among the best for information and meaningful activities to help children appreciate the importance and fragility of wetlands.

TIERNEY, ROBERT J., et al. 1995. *Reading Strategies and Practices*. Boston: Allyn and Bacon.
Current practices to scaffold reading comprehension are offered with concise descriptions and comments on how best to employ the strategy. Various assessment techniques are also described.

TOPS LEARNING SYSTEMS. Canby, OR.
Practical and clever guides on many different topics with a variety of hands-on activities using common materials.

WHITIN, PHYLLIS, AND DAVID J. WHITIN. 1997. *Inquiry at the Window*. Portsmouth, NH: Heinemann.
An interesting look at how visiting birds can enhance inquiry across the curriculum.

Children's Literature and Resources

ALIKI. 1990. *Fossils Tell of Long Ago*. New York: HarperTrophy.

BJORK, CHRISTINA, AND LENA ANDERSON. 1992. *Linnea's Windowsill Garden*. Stockholm: Raben & Sjogren.
A practical and child-friendly guide to gardening with Linnea, a character named after the little pink woodland flower.

BRANLEY, FRANKLYN. 1988. *The Planets in Our Solar System*. New York: Harper.

BRENNER, BARBARA. 1970. *A Snake-Lovers Diary*. New York: Harper.

DRAKE, JANE, AND ANN LOVE. 1994. *Official Kids' Summer Handbook*. New York: Ticknor and Fields.
Contains plans for making sweep nets as well as many other ideas for outdoor explorations.

GEORGE, JEAN CRAIGHEAD. 1992. Moon Book Series. New York: HarperCollins.
Thirteen book series, each highlighting a different animal. Unique style of writing throughout the series is great for read-aloud and/or research.

———. 1993. *The Fire Bug Connection*. New York: HarperCollins.
Excellent ecological mystery read-aloud that models an inquiry approach to problem solving.

———. 1996. *Tarantula in My Purse*. New York: HarperCollins.
A collection of true stories told by Ms. George about the animals she and her family experienced and cared for through the years.

GEORGE, LINDSAY BARRETT. 1995. *In the Woods: Who's Been Here?* New York: Greenwillow.
Wonderful read-aloud which can involve the listeners.

HART, AVERY, AND PAUL MANTELL. 1996. *Kids' Garden*. Charlotte, VT: Williamson.
Includes many ideas to help get kids started with both indoor and outside gardens.

KNEIDEL, SALLY STANHOUSE. 1993. *Creepy Crawlers*. Golden, CO: Fulcrum.
Resource for devising and conducting experiments with crawling creatures.

KRAMER, DAVID C. 1989. *Animals in the Classroom*. New York: Addison-Wesley.
Guide to caring for many of the most popular classroom animals.

KRAMER, STEPHEN P. 1987. *How to Think Like a Scientist*. New York: Crowell.
A wonderful guide for taking children through the processes of thinking scientifically.

MAZER, ANNE. 1991. *The Salamander Room*. New York: Dragonfly.
Great read-aloud.

McNULTY, FAITH. 1986. *The Lady and the Spider*. New York: Harper.
Popular read-aloud about a lady's encounter with a spider from her garden.

MILORD, SUSAN. 1989. *The Kids' Nature Book*. Charlotte, VT: Williamson.
Hundreds of ideas for nature-oriented activities.

PETERSON'S FIRST GUIDES. Boston: Houghton Mifflin.
These field guides are simple enough for most elementary students, yet packed with tons of information. Indispensable for going outdoors.

RANGER RICK NATURE SCOPE. Washington, D.C.: National Wildlife Federation.
This series contains terrific background for teachers on a variety of topics with effective activities for students.

SCHLICHTING, SANDI. 1985. *You, the Investigator*. Riverview, FL: Idea Factory.
A child's guide to selecting and planning a scientific investigation.

BERGER, MELVIN. 1995. *Germs Make Me Sick*. New York: Harper.

SIMON, SEYMOUR. 1975. *Pets in a Jar*. New York: Puffin.
Practical guide to finding and caring for some of the most common classroom animals.

Sources for Inquiry Materials

American Science and Surplus
3605 Howard St.
Skokie, IL 60076
Huge variety of inexpensive and hard-to-locate items.

Educational Innovations Inc.
151 River Rd.
Cos Cob, CT 06807
Ultraviolet detecting beads and other items suited for inquiry.

Edmund Scientific
101 East Gloucester Pike
Barrington, NJ 08007
General science items for the classroom.

Carolina Biological Supply
2700 York Rd.
Burlington, NC 27215

Acorn Naturalists
1730 East 17th St., #J-236
Tustin, CA 92780

Index